Jossey-Bass Teacher

*J*ossey-Bass Teacher provides educators with practical knowledge and tools to create a positive and lifelong impact on student learning. We offer classroom-tested and research-based teaching resources for a variety of grade levels and subject areas. Whether you are an aspiring, new, or veteran teacher, we want to help you make every teaching day your best.

From ready-to-use classroom activities to the latest teaching framework, our value-packed books provide insightful, practical, and comprehensive materials on the topics that matter most to K–12 teachers. We hope to become your trusted source for the best ideas from the most experienced and respected experts in the field.

Math Teacher's Survival Guide

Practical Strategies, Management Techniques, and Reproducibles for New and Experienced Teachers, Grades 5–12

JUDITH A. MUSCHLA

GARY ROBERT MUSCHLA

ERIN MUSCHLA

JOSSEY-BASS
A Wiley Imprint
www.josseybass.com

Published by Jossey-Bass
A Wiley Imprint
989 Market Street, San Francisco, CA 94103-1741—www.josseybass.com

ISBN 978-0-4704-0764-6

Printed in the United States of America
FIRST EDITION

PB Printing   10 9 8 7 6 5 4 3 2 1

# About This Book

The demands on math teachers have never been greater. Not only must they plan and deliver fundamental math instruction that meets state standards and satisfies the needs of diverse learners, they must teach problem-solving and critical-thinking skills, integrate technology in student learning, and prepare their students for standardized tests. Along with all this they must efficiently manage the other typical teacher responsibilities—grading papers, record keeping, handling behavior issues, and interacting with students, staff members, administrators, and parents and guardians. There never seems to be enough time in the school day to do all that must be done.

Teaching math, without question, can be stressful and frustrating, and balancing a career with personal and family life can be a constant struggle. Yet teaching math can also be a wonderfully satisfying and enjoyable career. Few things are as professionally rewarding as knowing you are working in an exciting and important field and helping to prepare students for a successful future.

The *Math Teacher's Survival Guide, Grades 5–12* is designed for the experienced and the new math teacher. The guide, which supports the Standards and Focal Points of the National Council of Teachers of Mathematics (NCTM), provides useful information and practical suggestions for making your classroom a dynamic center for student learning. Addressing essential topics from the start of the school year to its end, the guide will help you organize your teaching day, meet the daily challenges of providing effective instruction, and manage the routines that ensure learning in a classroom that celebrates responsibility, hard work, and respect for others.

Designed for easy use, the guide has several helpful features, including:

- A logical, step-by-step approach that offers helpful information on key topics
- A detailed Table of Contents that enables you to find topics quickly
- Cross-referencing of topics that allows you to easily access further information
- Numerous sources of additional information, including Web sites for both math teachers and students, as well as a list of References and Suggested Reading at the end of the book
- Reproducible management forms for teachers (also on CD and available at www.josseybass.com/go/mathteachersurvival.com).
- Reproducible information sheets for students (also on CD and available at www.josseybass.com/go/mathteachersurvival.com).
- A Quick Review at the end of each section that highlights major ideas and vital information

The *Math Teacher's Survival Guide, Grades 5–12* will help you not just to survive the daily demands of teaching, but will also help you to continue to grow professionally and become the best math teacher you can be. Our best wishes for a productive, successful, and rewarding year.

Judith A. Muschla
Gary Robert Muschla
Erin Muschla

# The Authors

*J*udith A. Muschla received her B.A. in Mathematics from Douglass College at Rutgers University and is certified to teach K–12. She taught mathematics in South River, New Jersey, for over twenty-five years. She taught at various levels at both South River High School and South River Middle School. In her capacity as a team leader at the middle school she helped revise the mathematics curriculum to reflect the Standards of the National Council of Teachers of Mathematics, coordinated interdisciplinary units, and conducted mathematics workshops for teachers and parents. She was a recipient of the 1990–1991 Governor's Teacher Recognition Program award in New Jersey, and was named 2002 South River Public School District Teacher of the Year. She has also been a member of the state Review Panel for New Jersey's Mathematics Core Curriculum Content Standards.

Together, Judith and Gary Muschla have coauthored several math books published by Jossey-Bass: *Hands-on Math Projects with Real-Life Applications, Grades 3–5* (2009); *The Math Teacher's Problem-a-Day, Grades 4–8* (2008); *Hands-on Math Projects with Real-Life Applications, Grades 6–12* (1996; second edition, 2006); *The Math Teacher's Book of Lists* (1995; second edition, 2005); *Math Games: 180 Reproducible Activities to Motivate, Excite, and Challenge Students, Grades 6–12* (2004); *Algebra Teacher's Activities Kit* (2003); *Math Smart! Over 220 Ready-to-Use Activities to Motivate and Challenge Students, Grades 6–12* (2002); *Geometry Teacher's Activities Kit* (2000); and *Math Starters! 5- to 10-Minute Activities to Make Kids Think, Grades 6–12* (1999).

Gary Robert Muschla received his B.A. and M.A.T. from Trenton State College and taught in Spotswood, New Jersey, for more than twenty-five years. He spent many of those years in the classroom teaching mathematics at the elementary school level. He has also taught reading and writing and is a successful author. He is a member of the Authors Guild and the National Writers Association.

In addition to math resources, he has written several resources for English and writing teachers, among them *Writing Workshop Survival Kit* (1993; second edition, 2005); *The Writing Teacher's Book of Lists* (1991; second edition, 2004); *Ready-to-Use Reading Proficiency Lessons and Activities, 10th Grade Level* (2003); *Ready-to-Use Reading Proficiency Lessons and Activities, 8th Grade Level* (2002); *Ready-to-Use Reading Proficiency Lessons and Activities, 4th Grade Level* (2002); *Reading Workshop Survival Kit* (1997); and *English Teacher's Great Books Activities Kit* (1994), all published by Jossey-Bass.

Erin Muschla received her B.S. and M.Ed. from The College of New Jersey and is certified K–8 in elementary education with Mathematics Specialization in Grades 5–8. She currently teaches seventh-grade math at Applegarth Middle School in Monroe, New Jersey.

# Acknowledgments

We thank Jeff Gorman, assistant superintendent, and Robert O'Donnell, mathematics and educational technology K–12 supervisor, of Monroe (New Jersey) Public Schools for their support of this project.

We also want to thank Chari Chanley, principal of Applegarth Middle School in Monroe, for her support.

Our thanks to Maria Steffero, math teacher at Applegarth, for her insightful comments and suggestions regarding topics in this book.

We wish to especially thank Kate Bradford, our editor at Jossey-Bass, for her guidance and advice from the initial idea for this book to its final production.

We also wish to thank Diane Turso for proofreading and helping to finalize this book.

And finally, we thank the many colleagues who have supported and encouraged us over the years, and the many students whom we have had the pleasure of teaching.

Contents

Section Eight: Building a Positive Environment for Learning Math ... 145

Section Nine: Interacting with Your Students ... 173

Section Thirteen: Managing Inappropriate Behavior 283

SECTION ONE

Embracing the Profession of Teacher of Mathematics

As a certified math teacher, you are a professional educator. You have completed the necessary courses, demonstrated proficiency in mathematics, and acquired a variety of teaching techniques to provide meaningful instruction to your students. But as significant as these accomplishments are, they constitute only a part of your responsibilities as an educator.

Your professionalism is founded on your beliefs, attitudes, and actions, and extends well beyond the classroom. For example, in addition to teaching, you must build your lessons around clearly stated objectives; support school policies and procedures; attend meetings, workshops, and conferences; serve on committees; interact with administrators, colleagues, students, and parents and guardians; and dress and conduct yourself with discretion and common sense. You must acquire and maintain good work habits, and constantly strive to develop your knowledge and expertise. In short, you must fully embrace the profession of teacher of mathematics and all that it requires in everything that you do.

Traits of Great Math Teachers

There are math teachers, and there are great math teachers. Math teachers become great math teachers through commitment, dedication, and enthusiasm. They work hard to develop their professionalism and share many of the following traits. Great math teachers:

- Understand the content of the courses they teach
- Use state and district standards and goals to plan and deliver instruction

- Utilize the Principles and Standards of the NCTM, and the Focal Points of the NCTM as important resources in developing their instructional programs
- Plan lessons that are based on the abilities and interests of their students
- Design lessons that will enable students to learn math skills and concepts
- Provide activities to meet the needs of students with various learning styles
- Act as a facilitator of learning
- Develop and maintain a practical set of classroom procedures and rules
- Foster a classroom atmosphere that promotes learning
- Develop and use a fair grading system
- Evaluate student progress consistently, both formally and informally, and provide regular and prompt feedback
- Teach and encourage the use of various problem-solving techniques
- Ask questions that require higher-level thinking and are relevant to their students' lives
- Provide problems that can be solved by a variety of methods
- Use technology in instruction
- Encourage their students to use technology to solve real-life problems
- Provide activities that promote cooperative learning
- Require students to write about and explain math concepts, problem-solving strategies, and solutions to problems
- Promote mathematical reasoning
- Encourage divergent thinking
- Use manipulatives and models to demonstrate math concepts
- Are receptive to new ideas and teaching strategies
- Are willing to collaborate with other math teachers for planning and instruction
- View math as a subject that all students, regardless of gender, ethnicity, or background, can learn
- Encourage all their students to do their best every day
- Are demanding in their expectations but are also considerate of their students' feelings and concerns
- Are consistent and fair
- Set realistic goals for themselves and their students
- Demonstrate the connection of math to other subject areas
- Are reflective and flexible

- Encourage students to apply math to their everyday lives
- Find genuine satisfaction in their students' growth

This seems like a lot, but we are sure that you can check off many of these things as already applying to your teaching. And with additional work, and the help of this book, you can acquire all of the professional traits that make math teachers great. One of your most important goals should be to become the best teacher you can.

Meeting State Standards and District Math Goals

A major aspect of your professional responsibilities is to ensure that your students meet or exceed the standards and goals established by your state and district. These objectives, which will help students attain the benchmarks of the No Child Left Behind Act, should be a part of your curriculum and be addressed in your daily instruction.

If you have not already done so, you can check the math standards of your state at www.educationworld.com/standards/state or by searching the Internet with the term "math content standards" and including the name of the state. You might also go directly to "math standards" on the Web site of your state department of education. Once you have obtained a copy of your state's math standards, keep it with your curriculum guide. You should refer to these standards as you plan your lessons, activities, and assessments. In addition to their standards, many state departments of education also include teaching guidelines, activities, and assessment materials to support teachers in their efforts to plan and deliver effective instruction.

To learn about any math goals your district has identified, check with your math supervisor or principal. In some school systems, district goals are revised yearly; in others, long-range goals may cover a few years. Incorporate district goals into your math lessons whenever possible.

The curriculum of every course contains a set of objectives which, together with state standards and district goals, provides a framework for the content of that course. Being aware of the major objectives and requirements of the math courses that precede and follow the courses you are teaching is vital information. Knowing what students have learned the previous year, what they need to learn to be successful in your class now, and what they will need to know to be successful next year helps you to plan instruction that will best meet their needs for long-term achievement in math.

Along with your curriculum and state standards and district goals, you should become familiar with the Principles and Standards of the National Council of Teachers of Mathematics (NCTM), as well as the Focal Points and Connections to the Focal Points of the NCTM, both of which can be obtained at www.nctm.org. These resources support a vision for proficient math instruction for all students and identify the math skills, concepts, and processes that students should master upon completion of specific grade levels. These resources can help you develop a successful and challenging math program.

In addition to understanding how state standards and district goals affect your teaching, you must also be aware of the prerequisites and requirements of each of your classes. Students who have not satisfactorily fulfilled the prerequisites for a course will likely experience difficulty in meeting the requirements of the course. A student cannot be expected to do well in Algebra II Honors if he has barely passed Algebra I. Although there are, of course, exceptions, overplacement is seldom beneficial to the student or his classmates. Students who are underplaced because they have exceeded the prerequisites of your course are also unlikely to benefit from it. You should consult with your math supervisor or guidance counselor to reassign incorrectly placed students to math classes appropriate for their abilities.

Understanding standards and goals enables you to provide instruction to your students that will help them learn the math concepts and skills necessary for them to satisfactorily complete your course. Standards and goals provide you with direction throughout the year.

School Policies and Procedures You Need to Know

Schools are complex institutions. For any school to function efficiently and safely, all staff members must understand the policies and procedures that govern its daily routines. Much of this information can be found in student and faculty handbooks, but some—especially revisions or additions to current practices—will be communicated during faculty meetings or via memos throughout the year. As a professional, you should know the policies and procedures for the following:

- Student attendance
- Homeroom procedures
- Tardiness
- Truancy
- Chronic absences
- Bell schedule
- Class schedule
- School calendar
- Signing in and out of school
- Earliest time faculty members can report to school
- Latest time faculty members may stay at school on a typical day
- Faculty attendance

- Faculty dress code
- Curriculum guides
- Unit plan format
- Daily lesson plan format
- Homework and classwork
- Grades
- Reporting periods
- Standardized testing schedule
- Teacher evaluations
- Substitute teacher plans
- Contacting substitute teachers
- Acceptable student behavior in class
- Acceptable student behavior in common areas, including outside the building
- Discipline
- Student fighting
- Harassment and bullying
- Detention
- Suspension
- Cheating
- Plagiarism
- Student dress code
- Referral of students for evaluation
- Individualized Education Programs (IEPs)
- 504 plans
- Bus plan for students
- Distribution of textbooks and other materials
- Lost textbooks and other materials
- Record keeping for books and materials
- Ordering supplies and materials from vendors
- Obtaining supplies and materials from the stock room
- Work orders for repair and maintenance of equipment
- Contacting the tech person in your school
- Contacting janitors
- Copy machine use

- Duty assignments
- Fire drill procedures for each class
- Emergencies and lockdowns
- School closings, early dismissals, and delayed openings
- Videotaping and photographing students
- Student lunches
- Teacher lunches
- Student use of the media center
- Student use of the computer lab
- Student use of technology
- Faculty meetings
- Faculty committees
- Faculty workshops and seminars
- Back-to-school night
- Contacting parents and guardians by phone and e-mail
- Parent conferences during conference time
- Parent conferences throughout the year
- Family vacations and student absences
- Extended student absence
- Student injuries in class or on school grounds
- Field trip procedures
- Taking students outside the building (for example, to do an activity on measurement)
- Classroom parties
- Food in the classroom
- Press releases
- Guest speakers
- Collecting money (for example, to pay for a field trip)

Once you understand the policies and procedures of your school, you must support and enforce them with fairness and consistency. Your students, colleagues, and administrators will respect your knowledge and dedication.

You also need to be aware of the chain of command in your school. Undoubtedly your school has procedures in place for management of discipline issues, referral of students to guidance counselors, and requests for the child study team to test and evaluate students for learning disabilities or emotional disorders. Following the correct procedures in such instances ensures that the proper people become involved and that they receive the necessary information for addressing the problem. By following your school's procedures, the issue has a greater chance of being resolved quickly and satisfactorily.

Only by understanding and supporting your school's policies and procedures can you assume your responsibilities in the daily program of your school. Knowing how and why things are done is an essential mark of a professional.

Professionalism and Common Sense

Just as mathematical knowledge, efficient classroom management, and effective instruction are critical components of a math teacher's professionalism, so is common sense. Sometimes, however, in the pressing demands of the school day, common sense can be overlooked. The consequences of ignoring common sense can be minor and slightly embarrassing—you are leaving school early and meet the superintendent on your way out—or they can be major and really embarrassing—you are talking in the faculty room about a student's continual lack of preparation, unaware that the substitute sitting next to you is his mother. Regardless of whether the outcome is minor or major, exhibiting a lack of common sense always undermines professionalism.

The following list clarifies instances and situations where common sense will help you to avoid making common (and not so common) mistakes:

- It is always better to arrive at school early. Use the time to grade a few papers, make copies, or respond to e-mail. Avoid arriving right on time, or worse, just a step ahead of your students.
- Leave school after the contracted time teachers may go. Even if you only stay a few extra minutes, you can update your assignments on your school's homework hotline or clean up papers on your desk.
- Adhere to the faculty dress code. If there is no dress code, wear clothing that you feel is appropriate. If you are not sure something is appropriate, do not wear it.
- Never use offensive language.
- Never discuss the behavior of students with other students.
- Never discuss a student with the parents or guardians of other students.
- Avoid gossip, which is often hurtful and is never professional.
- Never discuss students in the faculty room if substitutes or parents or guardians are present.
- Do not tell off-color jokes. Even though people may laugh to be polite, they may be offended.

- Never speak in a derogatory manner about any group.
- Do not speak negatively about another member of the staff or administration.
- Always give people your full attention when they are speaking to you. If they take the time to speak with you, you should take the time to listen.
- Do not talk to students as if they are your peers or friends. This does not mean that you cannot be friendly or informal at times, but always remember that you are their teacher. You must model professional behavior.
- Always attend required meetings and workshops.
- Be willing to serve on committees.
- Be attentive during faculty meetings, workshops, committee meetings, and seminars. Avoid grading papers, checking your cell phone for messages, or whispering to a colleague.
- Always follow school rules. Talking on your cell phone during class, for instance, sets a poor example for students who are not permitted to use their cell phones in school.
- Always be prepared for class. Lack of preparation shows students that the class is not important to you, and they may conclude it is not important for them either.
- Do not leave students in the classroom unattended. You are responsible for them and anything that may occur in class.
- Never lose your temper during meetings or conferences. Anger diverts energy from problem solving and makes finding solutions more difficult.
- Avoid being territorial when sharing rooms or supplies with other staff members.
- Do not hoard supplies.
- Do not monopolize the copy machine. If you have five hundred copies to run off, and a colleague has a handful, let her go ahead of you.
- Always complete paperwork on time.
- Never leave a classroom messy or in disarray for the teacher who has the room next. Just as you expect to enter a clean and orderly classroom, so do others. Be sure to leave the room before the other class begins.
- Never permit students to speak disrespectfully about other teachers, students, or classes.
- Always be tactful when speaking with parents or guardians, especially when you see them outside of school. Be discreet in what you say.
- Consider joining your parent-teacher association or similar organizations. Your membership and support will be appreciated.

- Volunteer your help to administrators, colleagues, and parent groups in your school whenever possible.
- Avoid making hasty decisions—they usually turn out to be the wrong decisions.
- Avoid procrastination. The more you procrastinate, the more work piles up, which will lead to frustration and stress.

When you combine common sense with sound teaching methods, you can become a role model for your students and for other teachers. All will see you as a professional who speaks and acts with intelligence, consideration, and good judgment.

Professionalism Outside the Classroom

Your students, their parents or guardians, and any other community members who know you are a teacher will view you as a teacher no matter where or when they see you. Certainly at any school function, such as parent-teacher association meetings, school fundraisers, or attendance at school sporting events—but even outside the school setting, such as at the mall, hairdresser, or place of worship—you need to present yourself as a professional member of your school's teaching staff. You must wear appropriate clothing, display proper conduct, and always use common sense. People will expect you to set a good example for their children.

Maintaining Your Professional Expertise

Acquiring and maintaining professional expertise is a goal you should pursue throughout your career. Only by constantly growing as a professional can you hope to provide the best learning environment and instruction for your students.

There are several ways you can improve your professional skills, including:

- Attend in-services, workshops, seminars, and conferences. Throughout the year most school districts offer in-services designed to foster the classroom management and instructional skills of their teachers. Many districts also provide money for teachers to attend out-of-district workshops, seminars, and conferences that present information on new techniques, strategies, or issues that can affect all aspects of teaching. Attending such events helps keep you current in trends in mathematical education and pedagogy and can inspire you with new ideas, activities, and methods.
- Further your own education by enrolling in graduate courses at local universities and colleges. Many school districts reimburse a portion of

tuition costs for graduate study. Before enrolling in any course, however, make sure that the course meets your district's guidelines for tuition reimbursement. An option is to enroll in courses that are offered online. To find online courses for math teachers, search with the term "math courses online for teachers."

○ Observe other math teachers. Seeing how your colleagues manage their classrooms and deliver instruction can offer valuable insight and give you ideas for improving your own methods. Before observing another teacher, always request his or her permission. Some people feel uncomfortable with another teacher in the room and you should respect their wishes. Other teachers will be happy to have you observe them. In this case, if possible, visit their classroom during a time that is convenient for them. You may later want to extend an invitation for them to sit in on your classes.

○ Join professional mathematics organizations. Such organizations keep you informed of current issues and trends affecting teachers, support your efforts in the classroom, and enable you to network with other teachers. The focus of these organizations vary: some concentrate on the needs of math educators or supervisors, and others address specific topics in the field of mathematics. Visit the Web sites of mathematics organizations to learn more about them. You might consider the following:

- American Mathematical Society (AMS), 201 Charles Street, Providence, RI 02904, www.ams.org. This society is for those individuals interested in mathematics and its application to everyday life.

- Association for Supervision and Curriculum Development (ASCD), 1703 North Beauregard Street, Alexandria, VA 22311, www.ascd.org. The ASCD is an organization for teachers and educational leaders.

- Association for Women in Mathematics (AWM), 11240 Waples Mill Road, Suite 200, Fairfax, VA 22030, www.awm-math.org. This association encourages women in mathematics and the sciences.

- Mathematical Association of America (MAA), 1529 18th Street, N.W., Washington, DC 20036-1385, www.maa.org. This association provides a forum for all those interested in mathematics.

- National Council of Supervisors of Mathematics (NCSM), 6000 East Evans Avenue, Suite 3–205, Denver, CO 80222, www.ncsmonline.org. This mathematics leadership organization provides information for school math supervisors and other educational leaders to enhance student achievement.

- National Council of Teachers of Mathematics (NCTM), 1906 Association Drive, Reston, VA 20191-1502, www.nctm.org. The NCTM is devoted to supporting the needs of math teachers.

- In addition to the above, you should check if your state has a professional organization for math teachers. Many do. Search the Internet using the term "professional math organizations" and include your state. Joining a state organization for math teachers provides you with the opportunity to attend workshops and conferences relatively close to home and meet with teachers from other school districts.

○ Join professional organizations for educators. Consider the following:

- American Federation of Teachers (AFT), 555 New Jersey Avenue, N.W., Washington, DC 20001, www.aft.org. The AFT is a teacher's union with 1.4 million members. It supports the interests of classroom teachers.

- National Education Association (NEA), 1201 16th Street, N.W., Washington, DC 20036-3290, www.nea.org. The NEA is the largest organization for public school teachers in the United States with close to 3.2 million members. It is a powerful advocate for public education.

- National High School Association (NHSA), 6615 East Pacific Coast Highway, Suite 120, Long Beach, CA 90803, www.nhsa.net. This association addresses the needs of high school educators.

- National Middle School Association, (NMSA), 4151 Executive Parkway, Suite 300, Westerville, OH 43081, www.nmsa.org. The NMSA is dedicated to the needs of middle school educators.

○ Subscribe to professional journals. Subscriptions to many journals are included when you join a professional organization. For example, the NCTM publishes *Mathematics Teacher* for math teachers of grades 8–14 and *Mathematics Teaching in the Middle School* for math teachers of grades 5–9. You might also consider the following:

- *Education Week,* Editorial Projects in Education, Inc., 6935 Arlington Road, Suite 100, Bethesda, MD 20814, www.edweek.org. Published weekly, this journal provides news and articles about education.

- *Instructor,* Scholastic, Inc., P.O. Box 713, New York, NY 10013, www.scholastic.com/instructor. For K–8 teachers, this resource offers practical articles on numerous topics and includes activities, teaching techniques, and reproducibles.

- *Teacher Magazine,* Editorial Projects in Education, Inc., 6935 Arlington Road, Suite 100, Bethesda, MD 20814, www.teachermagazine.org. This magazine provides teachers with information they need to provide quality instruction to their students.

○ Build a professional library. Start with the books and resource materials that you use for the courses you teach. Include your curriculum guides and the standards for your courses. Also include any texts your school no longer uses, which, even if they are dated, can be wonderful sources for ideas. A math dictionary and other math reference books, manuals for calculators or computer software, and faculty and student handbooks should also be a part of your library. You may expand your library by adding resource books, reproducibles, and materials for special activities.

⊙ Set yearly professional goals for yourself. As each year concludes, take some time to evaluate your performance as a teacher. Consider your strengths and weaknesses. Choose one or two areas in which you feel you could have done better and focus on improving these areas during the next year. Only concentrate on one or two, because attempting to work on too many will make your overall progress more difficult and likely frustrate you. You might consider the following:

- Improving your organizational skills
- Improving lesson planning for diverse learning styles
- Improving your skills in classroom management
- Incorporating technology in your lessons and student activities
- Implementing math projects in your program
- Integrating student writing in your curriculum
- Emphasizing problem-solving strategies
- Using portfolio assessment
- Improving your discipline
- Enrolling in a graduate program
- Handling paperwork more efficiently
- Creating rubrics for assessing student responses to open-ended questions

Achieving professional expertise as a math teacher is a significant accomplishment in your career. Continuing to grow as a professional is an even greater accomplishment.

Quick Review for Embracing the Profession of Teacher of Mathematics

Your professional obligations are apparent in all your interactions with your school community. Being aware of the following can help you to grow professionally throughout your teaching career:

⊙ Work to acquire the traits of great math teachers. These traits are the foundation of professionalism.

- Strive to meet state standards and district goals in your planning and instruction.
- Be knowledgeable and supportive of your school's policies and procedures.
- Always use common sense both inside and outside your school. Remember, you are a role model.
- Continue developing your professional expertise by:
  - Attending in-services, workshops, seminars, and conferences
  - Observing other math teachers and sharing ideas for teaching
  - Enrolling in graduate courses
  - Joining professional organizations, particularly those that specifically address the needs of math teachers
  - Subscribing to and reading professional journals
  - Building a professional library
  - Setting goals for your personal professional improvement

Your professionalism distinguishes you as a teacher. Arising from your commitment, dedication, and expertise, it inspires your students and everyone else in your school community to do the best they can in all they can. Your school is a better school because of you.

SECTION TWO

Before the First Day

*C*ongratulations! You are about to start a new school year.

If you are like most math teachers, you have plenty to do in preparation for meeting your students. Along with planning interesting and effective math lessons, you might need to implement a new curriculum, use a new textbook, or familiarize yourself with new technology. Maybe you will be working with new colleagues. Some of the supplies you ordered may not have arrived, you may have more students than desks in your classroom, and you may be unable to log on to your school's e-mail system. These are just some of the concerns you might have to address before classes begin, which is why the start of a new school year can be hectic and overwhelming.

The best way to manage the details that accompany the opening of school is to attend to them as quickly and effectively as possible. Resolving them prior to the first day enables you to concentrate on your most important priority: teaching.

Starting the Year Early

Whether you are a veteran or a first-year math teacher, early planning and preparation can lay the foundation for a successful and satisfying year. Going to school a few times in advance of the first day allows you to take care of matters that you may be hard-pressed to manage once school begins. It also gives you time to say hello to administrators, colleagues, and other staff members, meet with team members, and start to prepare for your students.

A day or two before going to school, call the main office to find out if your classroom will be available. During the break the maintenance staff will no doubt be making repairs, painting, washing and waxing floors, installing new equipment, moving furniture, and tending to countless other duties as they make your school ready for the coming term. A brief call enables you to schedule your visits on the days when you can get the most done.

Go to school with a list of goals you would like to accomplish that day. Be realistic with your expectations; trying to do too much will only frustrate you and sap your enthusiasm about the upcoming year. Remember to be flexible. If after arriving at school you find that other things need attention, adjust your plan. Always try to resolve the most important or pressing problems first. Smaller problems can then be solved in the time remaining before school or be addressed later when you have openings in your schedule.

The following reproducible, "Things to Do Before School Starts," contains items to address before the beginning of the school year. Use it as a checklist or as a guide to create your own checklist.

Things to Do Before School Starts

❑ Meet with administrators and your department supervisor in regard to the coming school year.

❑ Meet with team members, colleagues, and para-educators.

❑ Meet with support staff, including guidance counselors, and technology and media specialists.

❑ Obtain and review class lists.

❑ Obtain and review your schedule, room assignments, and any duties.

❑ Review your curriculum guides and teacher's editions of texts, especially if they have changed.

❑ Set up your classroom, including furniture, books, and supplies.

❑ Check equipment, such as calculators, computers, printers, projectors, and interactive whiteboards.

❑ Prepare materials for the first day, including information packets, descriptions of courses, and seating charts.

❑ Create lesson plans and math activities for the first day, or the first few days if possible.

❑ Set up your record book. Note the beginning and end of each marking period, and the days that school is not in session.

❑ Make copies before the copy machine rush.

❑ If you are a new teacher, find and meet with a mentor.

Depending on your situation, you may have other tasks you wish to accomplish before the opening bell. Managing them efficiently helps ensure a great start to your new school year.

Greeting Administrators, Colleagues, and Support Staff

When you visit your school before the start of the term, you should try to meet informally with administrators and those colleagues and staff members with whom you will be working. Along with a friendly hello and exchange of small talk, such meetings provide an opportunity to discuss new school policies, changes in curriculum or schedules, or any other news that may affect you and your students during the upcoming school year. Such conversations enable you to anticipate and prepare for changes.

During your visit, you should also introduce yourself to new personnel and offer to answer questions they may have. New staff members will appreciate your welcome and help. The relationships you establish with new staff members now will be the foundation for working with them later.

If one or more of your math classes is an inclusion class that receives in-class support from a special education teacher, or is a basic skills class that benefits from the presence of another teacher in the room, try to meet with these staff members before the beginning of school. If a face-to-face meeting is not possible, contact these individuals by phone. A major component of the success of any class in which two teachers are present is the relationship between them. Each of you has individual responsibilities regarding students, instruction, and classroom management, as well as responsibilities that you will share. You should discuss the students in the class who may need modifications, and you should discuss your roles, teaching methods that will best satisfy the needs of your students, and general expectations and procedures. (See "Working with Other Teachers" in Section Four.) You and your partner should strive to complement each other's classroom presence and teaching style. Deciding on roles, methods, and management prior to the start of school will result in effective procedures and routines from the first day of classes.

A successful school is built on the efforts of many people working together for the common purpose of educating children. Establishing and maintaining positive and professional relationships with administrators and other staff members should be one of your primary goals.

Getting a Head Start on Paperwork

The beginning of any school year is accompanied by an abundance of paperwork. Although some of the forms you will need to complete and sign may not be placed in your school mailbox until the students arrive, many may be waiting for you well in advance of the first day. In addition to paperwork related to your classes and students, you may need to update personal information. Some of your paperwork load at the beginning of the year may include the following:

- Updating your address, phone number, and e-mail for the school directory
- Notifying the individual in charge of personnel of a change in your name due to marital status
- Completing emergency contact forms
- Notifying administration about any additional courses you completed
- Providing transcripts or certifications for advanced credits or degrees
- Making changes in health care coverage
- Signing up for direct deposit of your payroll checks
- Acquiring a parking space and parking permit
- Reading and signing Individualized Education Programs (IEPs)

Paperwork that is not handled efficiently piles up and requires additional time to manage. A better option is to complete paperwork and hand it in as soon as possible, reducing the chances of required forms being late or lost.

Your Schedule and Class Lists

Two of the most important items you should obtain before the start of school are your schedule and class lists. Reviewing them now can prevent possible problems on the first day.

Check your schedule for the courses you are teaching, your room assignments, and any additional duties you may have. If you are teaching a new course or one you have not taught for a few years, you should review the curriculum, the teacher's edition of the text, the applicable math standards of your state, and the Standards of the National Council of Teachers of Mathematics (NCTM). Becoming familiar with the subject matter and course requirements will help you to plan effective lessons for your students.

In an ideal educational world, you would teach in only one room. If, however, you must travel between several rooms, consider the distance between them and how long it will take you to walk from one classroom to another through hallways clogged with students. If two of your classrooms are located on opposite ends of the school, think about asking your principal to change room assignments for you. Arriving at class after your students will delay the start of your lesson, and will undermine your rule for students to be on time for class.

As you review your schedule, note any duties that are assigned to you. Although duties are a part of just about every teacher's day, one that interferes with your arriving at your next class on time should be adjusted or changed.

When you look over your class lists, be sure to check the numbers of students in each class with the room assignment and the number of desks in each room. Starting the first day of classes without enough desks for students not only creates confusion and disrupts the assigning of seats, but it may also communicate a subtle message that the school does not care enough about its students to provide a desk for everyone. Even though this is surely not the case and the missing desks are in the room the next day, the effect of that first impression of math class may linger.

If your district provides you with additional lists of classified ESL (English as a Second Language) or LEP (Limited English Proficient) students, or students with other special needs, compare the names on these lists with your class lists. Make a special note of any students who have health issues, such as allergies, vision or hearing impairments, or physical handicaps. Being aware of these students can help you tailor your plans to meet the needs of everyone in your classes.

Seating Charts

Your class lists are the basis for your seating charts. For most classes, seating charts are useful and necessary. Most important, seating charts help you to learn the names of your students and convey to them that you are in charge of the class. Seating charts provide you with an orderly option of moving students to seats where they will benefit the most from your instruction, separate students who cannot get along (or who get along too well), and accommodate students with medical problems or physical disabilities.

Unless you know in advance of any special seating arrangements some students might need, you should make your first set of seating charts either randomly or in alphabetical order of students' last names. Either method works. As you come to know the strengths, weaknesses, and particular needs of your students, you can move them to different desks and revise your seating charts. If you prefer to use a traditional seating arrangement of rows, you may make copies of the following "Seating Chart Grid," or use the grid as an example to make a grid of your own. Other examples of seating arrangements are shown on "Seating Options." The arrangement in Option 1 is good for students to watch demonstrations with manipulatives, view screens, and use interactive whiteboards. The seating arrangement in Option 2 is excellent for group work, for it allows the teacher to circulate around the room and easily speak with each group. The arrangement for Option 3 fosters interaction between group members. Of course, you can use variations and combinations of these examples depending on your lesson. You might also prefer to keep your seating charts as electronic files on your computer and print out revised copies as necessary. Electronic grading programs often have the capability of creating seating charts. Yet another option is a magnetic seating chart that makes updating the chart easy when you change students' seats. Always keep updated seating charts for your substitute plans.

Seating Chart Grid

Subject _____ Room Number _____ Period _____

Seating Options

Option 1

Option 2

Option 3

Setting Up Your Classroom

Setting up your classroom requires a significant investment of effort and time. It is no small task to move furniture around, check equipment and materials, open boxes of supplies, and decorate your room with a mathematical theme that your students will find attractive and inviting. This work is best started and finished before the first day of school.

ARRANGING FURNITURE TO ENHANCE MATH LEARNING

The arrangement of the furniture and equipment in your classroom should facilitate your teaching of math. Whether you set students' desks in rows or in pairs or groups, all students should be able to clearly see you teach. They should have an unobstructed view of the board and any work placed on an interactive whiteboard, projector screen, or TV monitor. You may wish to ask your colleagues how they arrange the furniture in their classrooms. Though you do not have to do exactly what they do, their ideas may spark ideas of your own.

Before you arrange your room, you might find it useful to sketch a floor plan. Check your plan for practicality and safety. Students should have enough space to work comfortably, and there must be enough distance between desks and tables to move around the room easily. Not only do students need to be able to walk throughout the room, but you must be able to walk around the room so that you can meet with students individually or in small groups. Remember to safely secure any wires for electronic equipment. Wires laid across the floor present a tripping hazard. Any floor plan should support learning and be safe. Note: If you share a room with other teachers, consult them before making major changes to the room's layout.

As the year goes on, you may find that a different arrangement of furniture is beneficial to one or more of your classes. Do not hesitate to change. A practical yet comfortable plan for furniture in your classroom is an important step for creating a pleasant environment for learning.

CHECKING EQUIPMENT

Depending on your school and teaching situation, you may rely on a traditional chalkboard, overhead projector, and basic calculators for teaching math, or you may have access to the latest interactive whiteboard, digital projector, computers and printers, calculators, and calculator presenter. Whatever equipment you have for your classroom, you must make certain that it is in working order and that you know how to operate it.

If, at the end of the previous school year, you placed any work orders for equipment repairs or maintenance, check that these items have been returned to you and are functioning properly. Track down any equipment that is missing; resubmit work orders for any piece of equipment that is not operative. Once students arrive, you may not have time to search for or fix equipment.

Check the following equipment for your classroom:

- Computers and printers
- Projectors
- Enough calculators for your students (check that each calculator has functioning batteries)
- New batteries for calculators
- Installation of new software
- Cables
- Interfaces
- Power strips
- Extension cords (if needed)
- Tables or workstations
- Screens

If you need help in setting up any electronic equipment, contact the technical support person in your school. Be as specific as possible in describing the type of assistance you need. It will be easier to obtain assistance now rather than once school starts and just about everyone else needs help.

You should familiarize yourself with the operation of new equipment or software. An easy way to do this is to complete accompanying tutorials. If your school offers any workshops regarding new equipment or software programs, be sure to attend. Even if you feel that you will not have any trouble mastering a new device, the information you acquire at a workshop may prove to be helpful later. Always save any support materials and start-up or backup disks, and follow your school's procedures for registration of any warranties.

CHECKING MATERIALS AND SUPPLIES

Before school begins is also the time to check your materials and supplies. Starting classes without enough books, paper, or other necessary materials will undermine your efforts for beginning the year smoothly. Instead of becoming excited about what they will be learning in your math class, some students may be more concerned with not having received a book or the materials other students did.

In preparation for the first day, be sure to check the following:

- All the materials you will need for students in your homeroom (assuming, of course, that you have a homeroom). This could include: student manuals, insurance forms, parent or guardian emergency contact forms, health forms, and menus for lunch. (See "If You Have a Homeroom" in Section Six.)

- The teacher's editions of your math texts; also check for teacher's editions and answer keys for workbooks, enrichment workbooks, and other resources.

- Student texts, workbooks, enrichment workbooks, and other materials. Make sure that you have enough copies of each and that each book or workbook is in good condition.

- Basic supplies, including paper, pencils, graph paper, transparencies, stapler, staples, markers, chalk, transparent tape, paper clips, erasers, and other items essential to your classroom.

- Miscellaneous forms such as passes and sign-out sheets. You may find it useful to make copies of the "Hall Passes" and the "Classroom Sign-Out Sheet" that follow.

- Any materials you ordered, for example, math posters, pencils, or math manipulatives.

You will, of course, need more materials throughout the year. For a complete list of materials, supplies, and equipment essential to math classes, see Section Three, "The Math Teacher's Tools of the Trade."

Hall Passes

Hall Pass   Date_____ Time_____

Name_____

To:
○ Principal   ○ Guidance
○ Assist. Principal ○ Media Center
○ Nurse   ○ Lavatory
○ Other_____

Signature_____

Hall Pass   Date_____ Time_____

Name_____

To:
○ Principal   ○ Guidance
○ Assist. Principal ○ Media Center
○ Nurse   ○ Lavatory
○ Other_____

Signature_____

Hall Pass   Date_____Time_____

Name_____

To:
○ Principal   ○ Guidance
○ Assist. Principal ○ Media Center
○ Nurse   ○ Lavatory
○ Other_____

Signature_____

Hall Pass   Date_____Time_____

Name_____

To:
○ Principal   ○ Guidance
○ Assist. Principal ○ Media Center
○ Nurse   ○ Lavatory
○ Other_____

Signature_____

Hall Pass   Date_____Time_____

Name_____

To:
○ Principal   ○ Guidance
○ Assist. Principal ○ Media Center
○ Nurse   ○ Lavatory
○ Other_____

Signature_____

Hall Pass   Date_____Time_____

Name_____

To:
○ Principal   ○ Guidance
○ Assist. Principal ○ Media Center
○ Nurse   ○ Lavatory
○ Other_____

Signature_____

Classroom Sign-Out Sheet

Teacher's Name_____

Your Name	Destination	Date	Time Out	Time In

DECORATING YOUR ROOM

If you have your own classroom, or you share your room with another math teacher, you have the freedom to turn it into a complete and attractive environment for learning math. However, if you share a classroom with a teacher of another subject, you must be flexible in placing your math-related materials in the room. Perhaps you can speak with the other teacher about dividing the bulletin board space and any display areas.

Following are some of the ways to decorate your classroom for math:

- Check the Internet using the search term "math bulletin boards." You will find numerous helpful Web sites.
- Create bulletin boards that reflect the theme of your first unit. This will inform your students of the first topic in math.
- Display math posters that reinforce concepts you will teach throughout the year. You can either create these on your own, or purchase them from a variety of catalogues. (See "Sources for Math Materials and Manipulatives" in Section Three.) Here are some topics for math posters:
  - Order of Operations
  - Rules of Properties
  - Steps for Solving Word Problems
  - Problem-Solving Strategies
  - Formulas
  - Types of Geometric Figures
  - Slopes of Lines
  - Fractals
  - Equivalences
  - Famous Mathematicians
- Hang up posters showing class rules and procedures.
- Create models of two-dimensional and three-dimensional figures.
- Create bulletin boards of school news or a school calendar, especially one highlighting events in math.

Throughout the year, you may wish to change your posters or displays to support your current topic of study. Strive to create a classroom that reflects and celebrates mathematics.

Preparing for the First Day

In the days you spend at school prior to the arrival of students, you will be busy meeting with people, setting up your room, reviewing your curriculum and teacher's editions of your texts, and checking equipment and materials. But there is still more to do. You should now consider setting goals for your students, designing a handout of the rules and requirements of your classes, creating math activities for the first day, making copies of materials to be distributed to students, and checking that you can access your school's communications systems.

SETTING GOALS

The goals you set for your students should reflect your state's math standards, the Standards of the NCTM, and your district's guidelines, all of which are likely addressed in your curriculum. (See "Meeting State Standards and District Math Goals" in Section One.)

Along with the content goals stated in your curriculums, you should set process goals for your students such as the following:

- Students will become competent problem solvers.
- Students will use various strategies in problem solving.
- Students will clearly communicate mathematical ideas.
- Students will write explanations that detail mathematical reasoning in problem solving.
- Students will work together cooperatively.
- Students will become competent in the use of technology in math class.
- Students will come to appreciate the importance of math and apply math to their everyday world.
- Students will apply mathematics to other subject areas.

Clear goals provide you with purpose and direction, helping you to focus your planning and teaching. Knowing the destination always makes a journey easier.

RULES AND REQUIREMENTS OF YOUR CLASSES

A written summary of the rules and requirements of your classes provides your students with a blueprint for success in those classes. Such summaries also can alleviate students' fears regarding your requirements and their responsibilities.

Handing out a summary of rules and requirements for each of your classes on the first day is a practical way to introduce the class and emphasize important expectations. You might want to include the following in any class summary:

- Welcome and Introduction to the Class
- The Work You Can Expect
- Required Materials
- Classroom Rules
- Grading System
- Absences and Makeup Work
- Extra Help
- Special Notes

Try to limit your summary to two pages, and copy it on both sides of a single sheet of paper to eliminate the need for stapling. Instruct your students to place the summary in their binders. Copying the summary on colored paper will distinguish it from other sheets of paper in their binders. To create a summary of your own, refer to the following sample, "Rules and Requirements of Ms. Smith's Algebra I Class."

Rules and Requirements of Ms. Smith's Algebra I Class

Welcome to my Algebra I class. We will be studying expressions, functions, patterns, equations, and inequalities. You will learn to organize, analyze, and interpret data. You will also learn a variety of strategies that you will use to solve problems. The skills and concepts you learn this year will prepare you for advanced work in other math courses and science.

The Work You Can Expect

You will receive homework most nights except Fridays and before holidays. You can also expect classwork and group work. You will maintain a math notebook and journal. At least four quizzes will be given each marking period. Tests will be given after every chapter and unit. One project will be assigned during each marking period. There will be a midterm exam at the end of the second marking period and a final exam at the end of the year.

Required Materials for Class Each Day

- Three-ringed one-subject binder or a large binder with a section for math
- Assignment pad
- Algebra textbook
- Pencils with erasers
- TI-84 Plus Graphing Calculator with your name written on the back. Write your name in white correction fluid or nail polish or use a label maker. Keep the manual at your home for reference. If you provide me with the serial number of your calculator, I will keep a record of it.
- A black-and-white composition book for journal entries

Classroom Rules

- Come to class on time each day with the required materials and be ready to work.
- Respect yourself, others, and property.

(Continued)

- Be responsible for your actions.
- Follow directions.
- Raise your hand for a question or to make a comment.

Grading System

- Tests, 30%
- Quizzes, 20%
- Homework, 15%
- Classwork, 15%
- Projects, 10%
- Math notebook, 10%

Absences and Makeup Work

You are expected to make up all missed work because of an absence. If possible, check with a classmate to find out what assignments you missed. If you have any questions about the work, ask me. See me to make arrangements for making up any quizzes or tests. If you know that you will be absent, let me know ahead of time.

Extra Help

I will be in the library during Period 8 if you need extra help. I am also available before school and after school. Check with me one day before to make sure that I am free. Sometimes I must attend meetings during those times. You may e-mail me at school at jsmith@amiddleschool.org. I can be reached at school by phone at 123-456-7890, ext. 121.

Special Notes

Success in this class depends on hard work. You can do well if you try. I am looking forward to working with you this year.

Best wishes,

Ms. Smith

ACTIVITIES FOR THE FIRST DAY

For most teachers, the first day of classes can be described as having too much to do in too little time. There is the administrative work that must be done—taking attendance, sorting through scheduling mix-ups, assigning seats, distributing what may seem to be an endless stream of forms and papers—and, of course, you must start your math classes in an efficient manner.

Few math teachers run out of things to do on the first day. Most have trouble trying to fit everything in. After attending to the administrative details, you must carefully select those activities that will ensure a smooth start to the school year. The activities you choose to present to your students are largely dependent on time. If the first day of classes at your school is a half day, and the meeting time of your classes is shortened, you cannot do as much as teachers who teach in a block schedule or who have a full period.

Activities for your first day of classes might include some or all of the following:

- Welcoming students and introducing yourself
- Taking attendance
- Assigning seats
- Distributing required administrative handouts
- Providing a description of the class
- Discussing class routines
- Distributing a summary of the rules and requirements of the class
- Presenting an activity to help you get to know your students
- Presenting a math activity to emphasize the importance and relevance of math
- Distributing math texts, workbooks, and other materials

Planning activities for the first day well in advance enables you to select those that will best launch your math program. For more information on activities for the first day, see Section Six, "Planning a Great First Day."

MAKING COPIES BEFORE THE COPY MACHINE CRUNCH

If there is anything in any school that consistently breaks down at the worst possible times, it is the copy machine. To avoid long lines at your school's copier as the first day of classes approaches or, even worse, having to wait for a technician to arrive to repair the copier as students walk toward the front door, plan to make all of the copies you need for the beginning of school at least a few days before school officially begins.

BEING CONNECTED: CHECKING COMMUNICATIONS SYSTEMS

Schools of the twenty-first century require efficient communications systems. E-mail and voice mail are essential for communicating with colleagues, administrators, and parents and guardians. School Web sites, particularly homework links, enable students and parents and guardians to access information about assignments, and links to online grade books allow parents and guardians to regularly review the grades and progress of their children. If any of these systems is not functioning properly, your program will be undermined. Checking that your access to each is operational prior to the start of school not only reduces the chances of your missing important information in the days leading up to the opening of school, but can ensure that problems are uncovered and fixed before students arrive.

Your school may have several systems for communication. You should check that all to which you have access are functioning properly. Some of these systems may include the following:

- E-mail at your school address. You should be able to receive and send messages. (We suggest that you do not give your personal e-mail address to parents, guardians, or students.)
- Voice mail. Your name should be in the system. Your voice mailbox should save messages, and you should be able to retrieve those messages.
- Homework hotline. If your school has a homework hotline where you can leave a voice message of each night's assignment, you should make sure that you are able to record your message and that it plays correctly when accessed.
- Web site homework link. If your school's Web site has a link to an assignment page, be sure that you can post assignments.
- Web site grade book link for parents and guardians. Some schools maintain Web sites that parents and guardians can visit to check the grades of their children. If your school has this type of site, make certain that your grade book is set up correctly and that the proper safeguards are in place. Parents and guardians should only be able to view the grades of their own children.

The communications systems at your school are critical for sharing information and keeping you connected to others in your school community. If any of the systems is not working for you, contact the appropriate technical support personnel. Once school starts these individuals will quickly become backlogged with work.

Preparing for the first day in advance enables you to plan more effectively and reduces the chances for having to rush to attend to final details. You can enjoy the free time in the remaining days before school begins, knowing that you have done your homework, and look forward to the start of a new year in this demanding yet rewarding profession.

Especially for the First-Year Math Teacher

As a new math teacher, you are likely to be anticipating the start of the school year with a variety of emotions. Certainly you are looking forward to beginning your new profession. You are looking forward to working with your students and colleagues. You are looking forward to helping students gain the math skills that will help them realize success in the coming years. Yet you are somewhat anxious about the great responsibilities you will assume in fulfilling the demands of teaching. You may feel apprehensive regarding concerns about managing your classes; working with students, colleagues, and parents and guardians; fitting in at a new school; and doing and saying the right things.

Realizing and accepting the following facts can reduce first-year worries:

- Every new teacher becomes nervous as the school year approaches.
- Even veteran teachers experience some anxiety as they anticipate the start of a new school year.
- Thorough preparation can build your confidence and make apprehension more manageable.

As a new math teacher, you are much like the "new kid on the block." Your colleagues possess a knowledge of the workings of your school that you cannot yet understand. Whereas they easily settle into established routines in preparation for the approaching school year, you strive to prepare for things that, because of inexperience, you can only anticipate. Fortunately, you can start making the transition from beginner to veteran teacher from the moment you step into your school by learning as much as you can about your school, attending orientations, reviewing school handbooks, and teaming up with a mentor.

LEARNING THE LAYOUT OF YOUR SCHOOL

Your school may be small, housing only a few hundred students, or it may be a sprawling complex with several wings and numerous classrooms. Whether it is small or large, the fact that you do not know your way around will add to your worries about starting the year. Knowing the layout of your school will increase your self-confidence, for you will know where you are going and the best way to get there. It will also enable you to give clear directions on the first day to students who are lost and ask you where a particular room or office is located.

The weeks preceding the beginning of school are an excellent time for you to find your way around the building. The freedom to explore is greatest when the halls are not filled with students. You should learn the floor plan of your school as soon as you can. Most important, you should know where the school's main office is, as well as the offices of the principal, the vice principal, guidance counselors, nurse, and supervisor of the math department. In addition, you should know the location of the cafeteria, gym, media center, auditorium, copy machine room, and faculty lounge. As you learn about your school, you will meet other staff members. Be sure to introduce yourself. Being open and friendly will demonstrate to others your willingness to assume your place as a member of the staff. Do not underestimate the importance of first impressions. The relationships you form now have the potential to develop into friendships that will last throughout your career.

During your visits to school, go to each of your classrooms. Check the supplies and equipment in each room, and find where additional materials are stored. Familiarizing yourself with your work environment is an important step in becoming comfortable in your new position.

THE VALUE OF ORIENTATION

Many school districts offer an orientation for first-year teachers and for experienced teachers who are new to the district. Depending on the district, the orientation may range from an afternoon session to a week-long program. In some districts, teachers are compensated for attending a new-teacher orientation. In others, they are not. Regardless of compensation, you should attend any orientations that are offered. The goal of orientation is to acquaint new teachers with the procedures and policies of the district.

Though specific orientation programs vary, most are conducted by administrators, often with the assistance of veteran teachers. Many also include presentations by second-year teachers who benefited from a similar orientation the previous year. The observations and insights of these individuals can be quite helpful as they recently experienced much of what the district's new teachers will soon go through. At the orientation you may meet with various administrators, the school district's superintendent, some members of the board of education, experienced teachers, and teachers who are about to start their careers.

You will receive a significant amount of information at the orientation. Although much of the information will be in the form of overviews of your district's and school's mission statement, goals, and policies, administrators will also provide you with a summary of what you can expect from them and what they expect from you. If there is an opportunity to ask questions, do not hesitate to ask about things that concern you. It is likely that other new teachers have the same questions.

At some point during the orientation, a staff member or administrator may take the new teachers on a tour of the building, or perhaps the entire district. Even if you have already explored your building, you will probably learn something new. Your confidence will certainly increase as the school becomes more familiar.

Orientations for new teachers are a time to learn about your school and district. They also are a time to meet and interact with your future colleagues. You will find this a valuable learning experience that can help you achieve a smooth and positive start to the first day of school and your teaching career.

REVIEWING HANDBOOKS

Most schools compile information about policies and procedures in handbooks. Your school may provide handbooks for all faculty members, handbooks designed specifically for new teachers, and handbooks for students. You should review them all, for they contain information you need to fulfill your responsibilities as a staff member. (See "School Policies and Procedures You Need to Know" in Section One.)

FINDING A MENTOR

A mentor is a colleague who will provide one-to-one assistance and will support you throughout your first year as a math teacher. In some districts, a mentor will be assigned to you. In these districts, mentors are teachers who have been trained or who have demonstrated qualities that administrators feel are characteristic of someone who can help acclimate a new teacher to the school's methods, culture, and unwritten rules.

Over the years, mentoring programs have been established in schools throughout the country. More than thirty states have some form of mandated mentoring for new teachers. Sixteen states require and finance mentoring. In addition to districts with formal mentoring programs, many other districts provide some type of mentoring program. To find out if your state supports mentoring for new teachers, search the Internet with the term "mentoring new teachers" and include the state in which you will teach, or check your state department of education's Web site. You can also learn about the mentoring programs in specific school districts by searching the Internet with the term "mentoring new teachers" and including the name of the school district and state in which it is located.

A mentor is a valuable resource for a new teacher. If you are assigned a mentor, it is likely he or she will contact you. However, if you are not assigned a mentor, or your school district does not provide mentors, ask your principal or math supervisor to recommend a member of the math department who may be willing to work with you as an informal mentor. You could then contact this person and introduce yourself. If this individual is willing to work with you, set up a time to meet, preferably when both of you plan to visit the school before classes begin. Bring your class lists, schedule, and faculty and student handbooks to the meeting. Write a list of questions regarding the school, math curriculum, math resources, preparations for the first day, and any other issues or topics that concern you. Remember that your mentor is not a mind reader. He or she can only answer the questions that you ask.

Following is a list of sample questions that you might ask your mentor:

- Where do teachers park? Do I need a parking permit to park in the school lot? How do I obtain a permit?
- How early should I plan to arrive at school?
- Where do I report to school and do I need to sign in?
- Where do I obtain keys for my classrooms?
- How do I order a teacher's lunch? Are menus for teachers available?
- How do you arrange the furniture in your classroom?
- Will there be an abbreviated schedule for the first day of classes?
- Could you provide input on my plan for the first day of classes? Do you have any suggestions? What do you do on the first day?
- What are your class rules and requirements?
- What routines have you established for your students?
- I am unable to log on to the computer in my classroom. To whom should I report this? How do I contact this person?
- What type of lesson plans are required? Is there any specific format or template? How often should lesson plans be submitted? To whom do I submit them?
- How do you pace your lessons and units?
- Do most students have access to a computer at home?

A mentor can often be the difference between a first-year teacher floundering or flourishing. A mentor can help you take your first steps to what will be a successful career. (See "Responsibilities of a Mentor" and "Responsibilities of a Mentee" in Section Four.)

Every teacher was once a new teacher. Every veteran teacher in your school completed his or her first day and first year. You will, too.

Quick Review for Before the First Day

Beginning your preparations for a new school year prior to the first day gives you the opportunity to make certain that you will be ready for your students. Attending to the following will help you start your first day of classes smoothly and efficiently.

- Write a list of things you want to accomplish before school begins. Take the list to school with you.

- Say hello to administrators, colleagues, and members of the support staff. This is an excellent time to talk about changes in curriculum, schedules, or school policies that may have an impact on you and your students.

- Complete as much of the beginning-of-the-year paperwork as possible. The more you do now, the less you will need to do after school starts.

- Review your schedule and class lists. Consider any potential problems and take steps to prevent them.

- Review your curriculum, teacher's editions of your texts, state and district goals and standards, and the Standards of the NCTM that apply to the subjects you are teaching.

- Set content and process goals for your students.

- Check each of the classrooms in which you teach. Every classroom should be large enough and have desks for all your students.

- Create seating charts.

- Set up each classroom. Arrange the furniture; check to make sure equipment is working; check materials and supplies; and decorate the classroom for math.

- Write a summary of the rules and requirements of your classes for your students.

- Prepare activities for the first day of classes.

- Make copies of all the materials you will distribute to students on the first day, or, even better, the first few days.

- Check your communications systems at school, especially your e-mail account and voice mail. All systems should be functioning properly.

- If you are a first-year math teacher, learn about your school, attend any orientations or workshops for new teachers in your district, review any handbooks for teachers and students, and identify and meet with a mentor.

By getting ready for the new school year early, you allow yourself the time to prepare fully before your students arrive. Thanks to thorough preparation you will begin your first day of classes with confidence and effectiveness.

SECTION THREE

The Math Teacher's Tools of the Trade

Teaching mathematics is demanding and challenging. Each day you must provide interesting lessons that motivate students who have diverse learning styles and varying abilities. Hard work alone is not enough to accomplish this task. You also need the proper supplies, materials, equipment, technology, and resources. You need the math teacher's tools of the trade.

Basic Supplies, Materials, and Equipment

Although items such as pencils, pens, and paper are common supplies for every classroom, much of what you require is specific to math classes. Having everything you need to teach effectively frees you from having to search for an item before you can present a lesson, and enables you to devote more energy to teaching.

The following list contains the basics:

Pencils	Glue
Pens (black, blue, red)	Pencil erasers
Lined paper	Chalk
Construction paper	Board erasers
Grid paper	Index cards (various sizes)
Graph paper (various types)	Correction fluid
Poster paper	Poster mounting putty
Colored pencils	File folders
Rulers (metric and customary units)	Portfolios

(Continued)

Meter sticks	Storage bins
Yardsticks	Hole punch
Markers	String
Protractors	Self-adhesive note pads
Compasses	Projectors (digital or overhead)
Stapler (staples)	Transparencies
Pencil sharpener	Markers for transparencies
Paper clips (assorted sizes)	Screen or whiteboard for projector
Transparent tape	TV monitor
Calendar	Calculators (batteries and accessories)
Masking tape	Manuals for calculators
Tissues	Computers with Internet access
Scissors	Computer for teacher use only
Teacher's edition of texts	Manuals for computers
Copy of math curriculum	Math software
Copy of math standards	Manuals for software
Math dictionary and reference books	Interactive whiteboard
Teacher and student handbooks	Printer (ink cartridges)
Paper towels	Printer paper
Cleaning spray	Blank CDs, DVDs, and flash drives
Latex gloves	Storage containers for CDs and DVDs

Common Math Manipulatives

In addition to the basics, the following manipulatives can greatly enhance your instruction by allowing you to demonstrate properties or involve your students with hands-on activities for exploring concepts and relationships. A brief description of each manipulative and how it may be used is included.

- *Algebra tiles:* A set of positive tiles and negative tiles that represent units, x, and x^2. They are used to introduce operations and concepts of polynomials.

- *Attribute blocks:* A set of plastic blocks that includes circles, squares, rectangles, triangles, and hexagons. The blocks are used to classify shapes.

- *Base 10 blocks:* A set of cubes that represents 1, 10, 100, and 1,000. The cubes are used for developing an understanding of the decimal system.

- *Bills and coins:* Paper bills and plastic coins in various monetary denominations. They are used to demonstrate computation with money, including making change.

- *Centimeter cubes:* A set of several one-centimeter cubes. Each cube has a volume of one cubic centimeter and weighs one gram. They are used to explore surface area, volume, and equivalent measures.

○ *Clinometer:* An instrument that resembles a protractor attached to a viewing eyepiece. It is used to measure angles of elevation and find heights of large objects.

○ *Color tiles:* One-inch square tiles, usually available in red, blue, yellow, and green. They are used in generating and completing patterns and finding the probability of an event.

○ *Cuisenaire® Rods:* A set of ten rectangular rods in various lengths, ranging from 0.1 unit, 0.2 unit, and 0.3 unit to 1 unit. They are used in modeling the basic operations, computing with fractions and decimals, and enhancing spatial problem-solving skills.

○ *Decimal Squares®:* A set of small squares that represent a whole unit, 0.1, 0.01, and 0.001. The squares are used in exploring operations with decimals and percents.

○ *Dice:* Regular dice, with a number ranging from 1 to 6 expressed in dot notation. They are used in probability, especially experimental probability.

○ *Four-pan algebra balance:* A balance that comes with four plastic pans, weights, and canisters. It is used for deriving rules for working with signed numbers, solving linear equations, and modeling systems of linear equations.

○ *Fraction circles:* A set of several circles (the number varies according to the manufacturer). One circle represents a whole. The other circles are divided into parts, for example, halves, thirds, fourths, sixths, and eighths. They are used to model fractions, equivalences, and operations with fractions.

○ *Fraction squares:* A set of several squares (the number varies according to the manufacturer). One square represents a whole. The other squares are divided into parts, for example, halves, thirds, fourths, sixths, and eighths. They are used to model fractions, equivalences, and operations with fractions.

○ *Geoboards and rubber bands:* A square board with pegs on which rubber bands can be placed. The pegs may be set in a circular or square pattern. Circular geoboards can be used to discover properties of circles and congruence. Square geoboards can be used in graphing, finding area, and illustrating the Pythagorean Theorem.

○ *GEOFIX®:* A variety of shapes that students may hinge together to build three-dimensional models. They are used to demonstrate volume, perimeter, vertices, angles, and Euler's Formula.

○ *Geometric solids:* Wooden or plastic models of geometric figures including prisms, pyramids, cylinders, and spheres. They are used to help students visualize vertices, faces, and edges of polyhedra, and also to explore surface area and volume.

○ *Geo strips:* Plastic flexible strips of various sizes and connectors. They are used to create shapes and explore properties of polygons.

○ *Liter cube:* A 1,000-milliliter cube, graduated in 100-milliliter increments. It is used to measure capacity and demonstrate equivalent units.

○ *MIRA™:* A transparent geometric tool with reflective properties. It is used to introduce the concepts of symmetry and congruence, and to explore transformations.

○ *Mirror:* Any type of nonbreakable mirror can be used to explore reflections.

○ *Multilink® cubes:* Cubes of various colors that snap together. They are used for exploring cubes and squares of numbers, visualizing three-dimensional figures, and finding surface area and volume.

○ *Operations dice:* Six-sided dice that include $+$, $-$, $\times$, $\div$, $=$, and $>$ to practice math facts.

○ *Pattern blocks:* A set of hexagons, squares, trapezoids, triangles, and parallelograms. They are used to investigate patterns, congruence, and symmetry.

○ *Patty paper:* Translucent paper used to trace figures and investigate congruence, symmetry, and transformations.

○ *Pentominoes:* Twelve shapes scored in one-inch-square sections. They are used to explore area, perimeter, symmetry, congruence, and transformations.

○ *Playing cards:* A deck of fifty-two cards used to find the probability of an event.

○ *Polyhedra dice:* Unlike regular dice, each of these dice are four-, six-, eight-, ten-, twelve-, and twenty-sided, and are labeled with numbers ranging from one to the number of sides. They are used to investigate probability.

○ *Positive and negative number dice:* A set of eight cubes that are used for practicing operations with integers.

○ *Spinners:* Available in a variety of sizes and shapes, each spinner has an arrow that can be spun and then lands on a specific section of the base. Spinners are used to find the probability of an event.

○ *Tangrams:* A set of seven geometric figures: two large isosceles triangles, two small isosceles triangles, one parallelogram, one square, and one isosceles triangle that is not congruent to either of the other isosceles triangles. They are used to develop spatial sense and to explore congruence, area, and perimeter.

○ *Thermometer:* A Celsius/Fahrenheit thermometer can be used to show similarities between the two temperature scales. It can also be used to derive one of the conversion formulas from the other.

○ *Timer:* A digital timer. It is used to measure elapsed time or to record time in experiments.

○ *Trundle wheel:* A circular piece of wood or plastic that is attached to a long stick so that the circle may rotate. It is used to measure distances that are too long to measure with a meter stick or a tape measure.

○ *Two-color counters:* A set of circular disks. One side of each disk is red and the other side is yellow. They are used with probability, estimation, and modeling operations with integers.

○ *Unifix® cubes:* Interlocking cubes that link in only one way. They are used to create and extend patterns, and model fractions and decimals.

The manipulatives described above can be important assets to your instruction. Some manipulatives, including pattern blocks, tangrams, and decimal squares, are also available for use with overhead projectors. A search of the Internet using the term "math manipulatives for the overhead projector" will result in an abundance of visual aids you can use to demonstrate math concepts to students.

If you are unable to obtain all of the manipulatives you need for your program, perhaps you can obtain some and colleagues can obtain others. You may then share them as needed, or you may be able to make some manipulatives. To find templates or instructions for making manipulatives, search the Internet with the term "how to make . . ." and include the name of the item. (See "Handling Math Manipulatives Effectively" in Section Eleven.)

Sources for Math Materials and Manipulatives

Several companies specialize in providing materials and manipulatives for math teachers and their students. The following is a list of companies that will be able to meet your needs:

○ Didax Inc., 395 Main Street, Rowley, MA 01969, www.didax.com, 800–458–0024.

○ ETA/Cuisenaire, 500 Greenview Court, Vernon Hills, IL 60061, www.etacuisenaire.com, 800–445–5985.

○ Nasco, 901 Janesville Avenue, P.O. Box 901, Fort Atkinson, WI 53538-0901, www.enasco.com/math, 800–558–9595.

You may request a current catalog from each of these companies by visiting their Web site or contacting them by phone. There are, of course, many other companies that market math manipulatives. A search of the Internet using the term "suppliers of math manipulatives" will locate many more.

Technology

Technology is a significant aspect of your tools as a math teacher. Calculators and computers, along with a host of other devices to which they can be connected, support your teaching and expand the learning environment of your students. Ours is a digital age, and it is hard to imagine a modern math class without access to technology.

To gain the greatest benefits of technology, you must use these devices efficiently. Although most technology used in math classrooms is user-friendly, many teachers are intimidated by technology and implement it only at basic levels; doing so undermines the benefits that you and your students might otherwise gain. Reading manuals about hardware and software, going through tutorials, consulting with the tech person in your school, and attending workshops on technology can help you develop the confidence and expertise to incorporate the latest technological devices in your program.

CALCULATORS

Calculators are one of the most important tools in math classrooms for both teachers and students. You may use a calculator for computing students' grades; for checking the answers to problems you create; for following a student's reasoning as he explains how he arrived at the solution to an open-ended question; or, in conjunction with a calculator presenter, to project your graphing calculator's screen onto a classroom display to demonstrate the calculator's features.

Although most students view calculators primarily as a computational tool, calculators foster learning in various ways. By facilitating the exploration of mathematical patterns, and making it easier for students to formulate, test, and verify conjectures, calculators can help students develop higher-level thinking and problem-solving skills. They also make it easier for students to master mathematical concepts. For example, creating a list of the squares of numbers and another list of multiplying numbers by 2 reinforces the concept of squaring a number as opposed to multiplying a number by 2, thus helping students to avoid this common misconception. Capable use of calculators supports students in moving beyond basic math.

Calculators can be especially useful in motivating students by enabling them to experience success in math. Using calculators allows students to concentrate on finding the solution to problems without worrying about making a mistake in computation. This is a great benefit to students with weak computational skills who might otherwise be discouraged from attempting to solve challenging problems.

Because there are so many different models of calculators from which to choose, you should select a model with capabilities that are appropriate for your curriculum and your students' abilities. You can find information about a specific calculator by visiting its manufacturer's Web site. Along with tech support, there will likely be descriptions of accessories, software, and presentation tools, as well as downloads, resources, and activities for students. To find information about a particular calculator, search the Internet using the company's name. For example, if you are using a Texas Instruments calculator, search with the term "Texas Instruments calculators," then follow the appropriate link.

Whether you have chosen the calculator model for your students, or your school has purchased calculators for your math classes, you should learn how to use all of the calculator's features. Every calculator model has subtle differences, and possessing a sound understanding of the calculator your students are using enables you to teach its full capabilities. If the policy of your school is for students to purchase their own calculators, perhaps you can recommend the model they should consider. This helps to ensure that the calculators students purchase have the same features and are appropriate for their abilities. If every student is using the same calculator as you are, every student can use the same keystrokes to produce the same outcome. This makes it easier for you to demonstrate the features of the calculator. If a student uses a different calculator, the same keystrokes may produce different results, which may lead to confusion and frustration. When a student purchases a different calculator from the one the class is using, suggest that he retain the manual. Later, if he needs help, you may consult the manual and show him how to use his calculator.

Calculators are valuable tools in your math classes. Attaining skill in their use will promote the math literacy of your students and enable them to explore concepts and solve problems proficiently.

COMPUTERS

Computers serve many practical functions in math class. Your effective use of computers will enable you to broaden the scope of your overall program, enhance your instruction, and promote student learning.

You may use computers for:

- *Communicating* with colleagues, administrators, parents and guardians, and students via e-mail. E-mail is particularly useful for sending updates to parents and guardians informing them of their children's progress. It can also be used to send assignments or notes regarding missing work, upcoming tests or quizzes, or special events in your class. Some teachers encourage students and their parents and guardians to e-mail them at their school e-mail address with questions about assignments.

- *Writing.* You may create tests, quizzes, and worksheets. Using Microsoft Equations Editor, which is included in most versions of Microsoft Word (www.microsoft.com) or MathType by Design Science (www.dessci.com), you can create material that includes equations, fractions, exponents, mathematical symbols, and formulas. You can even insert geometric figures and graphs into your document to produce high quality work. Your students can use computers for writing reports on math topics or for explaining their solutions to open-ended problems.

- *Using an electronic grade book.* Most electronic grade books offer numerous features for recording, averaging, and posting grades. Most are also capable of generating seating charts, attendance reports, and missed assignments.

- *Developing a PowerPoint presentation* for students or for parents and guardians during back-to-school night.

- *Visiting Web sites and conducting research* on the Internet. While you may find ideas for developing lessons, using virtual manipulatives with your students, or downloading activities, students may find information for math reports or visit interactive math Web sites where they can explore concepts, view simulations, and play games that reinforce mathematical skills.

- *Presenting a lesson.* By linking your computer with a digital projector, you can display your computer's desktop onto a screen for your students to view. This is an excellent way to demonstrate virtual manipulatives and math concepts. You may also link your computer and projector to an interactive whiteboard where users can control computer applications using a finger.

- *Using software and special programs* to illustrate math concepts. For example, Geometer's Sketchpad (www.keypress.com) is an interactive program that students, or you, may use to investigate geometric and algebraic concepts and properties. Fathom Dynamic Data Software™ (www.keypress.com) allows students to model relationships, derive algebraic functions, and represent data in a variety of ways. Microsoft Excel can be used to create spreadsheets, from which tables, charts, and graphs can be generated. Handy Graph (www.handygraph.com) is software that produces Cartesian graphs and number lines. Many textbooks include software packages or access to the publisher's Web site that offers information and activities.

- *Creating a Web page* for your math classes. You may post due dates for projects and assessments, offer challenging problems for students to solve, or share math tips or trivia written by students. You may include a link for a parents' and guardians' page where you may provide information on upcoming tests, topics of study, activities, and other important announcements. To create a Web page for your classes, first check with the tech person at your school. She might be able to direct you to useful resources. You can also find information for setting up a Web page at Education World (www.schoolnotes.com) and Class Notes Online (www.classnotesonline .com).

- *Creating a Web quest* for your students. A Web quest is an activity that requires students to use the Internet to gather information to solve a problem. You may create a Web quest of your own, or search the Internet using the term "math Web quests." An excellent site for Web quests is Best Web Quests (www.bestwebquests.com).

When computers first appeared in math classes, most were used to present students with drills in the form of computational games. They were little more than supplementary tools. Since those early days, computers have increasingly been used for teaching concepts and higher-level skills and have become critical components of math programs.

INTERACTIVE WHITEBOARDS

An interactive whiteboard works in combination with a computer and projector. The board itself is an electronic, touch-sensitive display that has several uses in a math classroom, including:

- Presenting material on a computer screen to the entire class
- Working with and manipulating text, figures, and charts
- Viewing Web sites as a class
- Exploring virtual manipulatives on the screen
- Using or demonstrating the use of software on the screen
- Engaging students with the lesson by utilizing the touch-sensitive screen of the whiteboard

By incorporating an interactive whiteboard in a math lesson, you can provide computer-based learning on a large electronic display that allows great interaction for both you and your students. A search with the term "interactive whiteboards" will result in numerous informative and helpful Web sites. Three you might find particularly helpful include:

http://www.fsdb.k12.fl.us/rmc/tutorials/whiteboards.html
http://eduscapes.com/sessions/smartboard
http://education.smarttech.com

Resources on the Internet

The Internet is a near limitless source of ideas and support materials for your math lessons and instruction. With Web sites designed for teachers and students, the Internet is a vital tool for every math class.

Searching the Internet with the term "math Web sites for teachers" will yield so many sites that you could spend a good portion of your career visiting them. Fortunately, search engines list most of the best Web sites first; even better, we have included what we feel are some of the best sites for math teachers in the list below:

- About Mathematics: www.math.about.com. A wide assortment of information, including articles, lesson plans, and resources, is available on this site.

- Awesome Library: www.awesomelibrary.org. This Web site offers links to sites that include lesson plans, tutorials, and homework help on a variety of topics.

- Cut the Knot: www.cut-the-knot.org. This interactive site covers topics ranging from basic math through calculus, including fractals and combinations.

- Figure This!: www.figurethis.org/teacher_corner.htm. Various materials, resources, and handouts, including family activities, are included on this Web site.

- The Futures Channel: www.thefutureschannel.com. This site includes a library of online movies and lesson plans that allows students to explore various careers and real-life applications of math.

- High School Ace: www.highschoolace.com/ace/math.ech. Math tutorials and games are available on this site.

- Incompetech.com: www.incompetech.com/graphpaper. This Web site offers free online graph paper, including square grids, number lines, log/semi log, and polar graph paper.

- Math Bits: www.mathbits.com. This site offers algebra, geometry, and statistics lessons and activities for students in high school.

- Math.com: www.math.com. This Web site contains lesson plans, classroom resources, and standards, ranging from topics in basic math through statistics and calculus.

- Math Counts: www.mathcounts.org. The homepage of Math Counts contains problems of the week and also archived problems.

- Mathematics Archives: http://archives.math.utk.edu/calculus/crol.html. This Web site contains information and learning resources for teaching calculus.

- Mathematics Benchmarks: www.utdanacenter.org/k12mathbenchmarks. This site offers mathematics benchmarks for students in grades K–12, and also sample secondary assessments and tasks.

- The Math Forum: www.mathforum.org. Professional development, problems of the week, and math questions and answers are only a few of the components of this Web site.

- Math Motivation: www.mathmotivation.com. This site includes videos and lessons for high school math.

- Math World: www.mathworld.wolfram.com. This site offers an interactive math encyclopedia.

- Mega Mathematics: www.c3.lanl.gov/mega-math. Information on a variety of mathematical topics can be found here.

- National Library of Virtual Manipulatives: www.nlvm.usu.edu. At this site you will find virtual manipulatives for all grade levels in the strands of numbers and operations, algebra, geometry, measurement, data analysis and probability.

- The National Science Digital Library: www.NSDL.org. On this site teachers benefit from the free online library for educators that provides access to resources and tools for all levels of math education.

- NCTM Illuminations: http://illuminations.nctm.org. This Web site offers lessons and activities based on the NCTM standards. It also includes useful links to other sites.

- PBS Teachers: www.pbs.org/teachers. This site offers articles and resources for math teachers by grade levels.

- Purplemath: www.purplemath.com. This is an excellent algebra site that features online tutoring, quizzes, and worksheets.

- Shodor Interactivate: www.shodor.org/interactivate. This site provides activities with worksheets and lessons that introduce students to math concepts, vocabulary, and formulas.

- Soft Schools.com: www.softschools.com. Worksheets and games for middle school math are available on this site.

- Teachers TV: www.teachers.tv. This site offers free TV and online education programs.

In addition to Web sites geared specifically to math teachers, you will likely find these helpful as well:

- Education World: www.education-world.com
- The Educator's Reference Desk: www.eduref.org

- Federal Resources for Educational Excellence: www.free.ed.gov
- Internet 4 Classrooms: www.internet4classrooms.com
- Teachnology: www.teach-nology.com

Searching the Internet with the term "math Web sites for students" will identify numerous sites for students, including the following:

- Absurd Math Interactive: www.learningwave.com/abmath. On this site students use math skills to unravel the secrets of problem solving.
- Algebra Help: www.algebrahelp.com. This site provides support for algebra students.
- Coolmath.com: www.coolmath.com. Resources for algebra, geometry, fractions, fractals, and decimals are included on this site.
- Curious Math: www.curiousmath.com. This site offers math tricks and oddities.
- Fact Monster: www.factmonster.com. This Web site includes math games and facts for middle school students.
- Funbrain: www.funbrain.com/kidscenter.html. Students in grades 5 and 6 will enjoy a variety of interactive math games on this site.
- Funmaths: www.funmaths.com. This Web site offers interactive learning games for high school students.
- Hot Math: www.hotmath.com. On this site students will find step-by-step explanations for odd-numbered problems for 250 textbooks.
- The Math Forum: www.mathforum.org/dr.math. Students in all grades can ask Dr. Math questions about mathematics or read responses to students' questions.
- Math Is Fun: www.mathisfun.com. This Web site features games, puzzles, and activities about algebra, geometry, and measurement.
- Mathmistakes.info: www.mathmistakes.info. This Web site shows common errors in algebra, trigonometry, and calculus, and how to correct them.
- Maths Dictionary for Kids: www.amathsdictionaryforkids.com. This interactive site contains math definitions for elementary and middle school students.
- National Council for Education Statistics: www.nces.ed.gov/nceskids. On this Web site middle school students can create graphs and explore probability.

⊙ SpaceTime Arcade: www.spacetime.us/arcade. This site offers free mathematics games and puzzles.

The Web sites listed for teachers and students are just a start. Periodically spending a few minutes visiting math Web sites will give you an inexhaustible amount of ideas and activities to keep your classes fresh and exciting.

Sources of Supplementary Teaching Materials

Most publishers of math textbooks include extensive support materials to be used with their programs. Depending on the program your school purchases, along with your teacher's edition and the texts for your students, you might also have student workbooks, supplementary or enrichment activities, suggested projects, CDs or DVDs, and online access to the publisher's Web site where you can find additional material for yourself and your students. Yet, you may still require more new ideas and fresh materials. For those times, you might consider the books and resources offered by the publishers that follow:

⊙ Continental Press: www.continentalpress.com

⊙ Creative Publications: www.creativepublications.com

⊙ Eye on Education: www.eyeoneducation.com

⊙ Heinemann: www.heinemann.com

⊙ Incentive Publications: www.incentivepublications.com

⊙ Jossey-Bass: www.josseybass.com

⊙ J. Weston Walch: www.walch.com

⊙ Key Curriculum Press: www.keypress.com

⊙ Lorenz Educational Press: www.teachinglearning.com

⊙ McGraw Hill: www.mheducation.com

⊙ National Council of Teachers of Mathematics: www.nctm.org

⊙ Prufrock Press: www.prufrock.com

⊙ Scholastic: www.scholastic.com

There are many other publishers of materials for math teachers. A simple search of the Internet with the term "educational publishers, math" will result in many more.

Quick Review for the Math Teacher's Tools of the Trade

The days of teaching math with chalk and chalkboard and pencil and paper have passed. Math teachers today rely on what their predecessors of just a generation ago would consider to be an astonishing assortment of tools. The following highlights your tools of the trade:

- Obtain all of the supplies, materials, and equipment you need to teach your students effectively. This includes everything from pencils and paper to calculators and computers. Lacking anything you need makes your teaching day more demanding and detracts from the learning environment.

- Acquire math manipulatives with which you can demonstrate mathematical properties and concepts to your students. Encourage your students to use manipulatives to explore mathematical relationships and model ideas.

- Implement technology in your classroom. Use calculators and computers to broaden the scope of your instruction and enrich the learning experience of your students. Incorporate technology into your curriculum so that it furthers your instructional objectives.

- Use the Internet as a source of ideas for enhancing your instruction and also as a resource for your students. Countless Web sites offer information, activities, and insights about mathematics.

- Supplement your program with resources that motivate and challenge your students.

With the proper supplies, materials, equipment, technology, and resources—your tools of the trade—you are better able to plan and provide interesting and effective lessons for your students. This is a major step toward building a quality math program.

SECTION FOUR

Becoming a Valued Member of the Staff

*S*chools are complex work environments. Although much of your typical day will be spent working with students and other teachers, you may also work with administrators, guidance counselors, members of the child study team, para-educators, secretaries, custodians, substitute teachers, and even, on occasion, security personnel. All of these individuals work in common purpose for the education and welfare of the students in your school.

Cooperation among staff members is critical for a school to function effectively. But because every individual has his or her own unique personality, sharing a common purpose is not always enough to ensure cooperation. Sometimes different personalities have difficulty working together. For example, a strict, very structured math teacher can quickly find herself in conflict with her in-class support special education teacher who feels that there are times when spontaneity can be a refreshing change. Even if these two individuals agree on what should be done, they may not agree on how it should be done. They may share the same objectives for helping their students master math skills, but they may not be able to coordinate their efforts. Instead they follow their own educational philosophies and expectations for student learning, resulting in an inconsistent instructional program that may prevent students from reaching their full potential.

The complexity of a school community guarantees that staff members must work together every day in a variety of situations. When people cooperate to find solutions to problems, usually the problems are solved more quickly. Individuals who are able to work effectively with others become valued members of their staffs.

Working Effectively with Others

Being able to work well with others is one of a teacher's most important abilities. It requires numerous skills that develop with time and experience. Though personality traits clearly affect how well people work together, the following suggestions can help you nurture your interpersonal and cooperative skills:

- Commit yourself to the best interests of your students and school.
- Treat others with courtesy and respect.
- Communicate your ideas clearly.
- Listen politely and with interest to the ideas of others.
- Be receptive to new ideas.
- Work toward mutual goals.
- Be willing to cooperate and compromise.
- Be an active participant in the decision-making process.
- Be enthusiastic in your responsibilities and duties.
- Be considerate of the opinions and feelings of others.
- Be dependable.
- Be trustworthy.
- Help others and acknowledge when others help you.
- Cheer the success of others.
- Do the extra things that result in success for all.

As you work with others to ensure the successful operation of your school, you will build trust. You will know that you can rely on your colleagues, and they will know that they can rely on you. Such credibility strengthens working relationships and is a critical factor for effective handling of a school's daily challenges.

THE CHAIN OF COMMAND IN YOUR SCHOOL

Every school has a chain of command, which is its structure of supervision. Everyone in a school system works under the supervision of someone else. Even the superintendent is accountable to the members of the board of education. Understanding and following the chain of command shows that you are a professional and are respectful of protocol.

Your chain of command, from top to bottom, is likely to be similar to the following:

- Board of education
- Superintendent of schools

- Assistant superintendent
- Curriculum supervisor
- Director of special services
- Principal (of your school)
- Assistant principal
- Math department chairperson or supervisor
- Team leader
- Teacher

Always follow your chain of command. Ignoring it may cause others to question your judgment and professionalism. Do not, for example, see your principal about a specific math course when your concerns can be better addressed by your math department chairperson. Not only will your department chairperson likely be displeased that you apparently do not respect his position enough to discuss a situation that falls within his duties, but your principal will likely be disappointed that you are taking her time with an issue that someone else has the expertise and responsibility to handle.

Although the greatest part of your day will be spent teaching, there will be times when you need to work with administrators and supervisors to solve problems. At these times be sure to follow the chain of command.

WORKING WITH OTHER TEACHERS

You are a member of the math department of your school or district, and you might also be a member of a team on your grade level. Perhaps you work in a team-teaching program or have special education teachers providing in-class support to one or more of your classes. Most math teachers work closely with other teachers every day.

If you are a member of a team, you not only have a responsibility to your students but to your team as well. You must support team goals, rules, and procedures. You may be involved in planning and implementing interdisciplinary units of study, and need to coordinate your lessons with your team so that math topics reach across the curriculum. Perhaps each member of your team has a special role, such as parent-guardian liaison, field trip director, or team Web site manager. You may also meet as a team with parents and guardians, in which case a unified approach is essential for resolution of potential problems. All this requires cooperation, flexibility, and compromise. Being a member of a team allows you to be a part of a small learning community within the larger school environment.

Sometimes another teacher may work with you in your classroom. An example might be a special education or basic skills teacher. In such cases, both of you must agree on how the class will be managed. You must discuss your needs and expectations, teaching styles, and instructional delivery systems. You must consider the role that each of you will assume in class. If a special education teacher is assigned to work in your classroom, you might consider team teaching (also called cooperative teaching). In this approach, you and the special education teacher share responsibility for specific classroom instruction. Together you would plan how to teach math to all of the students in class, with you being responsible for the entire class, and the special education teacher being responsible for the progress of the special education students, especially the implementation of their IEPs. If a basic skills teacher is assigned to your classroom, you may assume the responsibility for introducing and teaching the math lesson to the class, while the basic skills teacher focuses his efforts on providing support to students who need additional help. There are, of course, many other ways to divide responsibilities; however, any classroom management system you implement should be comfortable for all teachers involved and should provide instruction that maximizes the learning of all students in the class.

Whether you are working as a member of a team, in a team-teaching classroom, or simply as a member of your school's staff, your colleagues can be a source of support and encouragement. Working together can make the job easier for all.

WORKING WITH PARA-EDUCATORS

A para-educator (often referred to as a paraprofessional or an instructional aide) is an individual who works directly with special-needs students. Many para-educators assist special education teachers, and many others work with students in regular classrooms. They may work with several students, or they may be assigned to work with only one throughout the day.

If a para-educator is assigned to one or more of your students, you should involve this person in your class as much as possible. Underutilizing her skills ignores a valuable resource to you and the students. Discuss with the para-educator what her role will be in your classroom, and work together to establish her responsibilities and duties. Note, however, that depending upon her certification, a para-educator may not be able to assume certain types of responsibilities. Most para-educators are not legally permitted to be left alone in the classroom to supervise students.

Because para-educators work closely with the students to whom they are assigned, their understanding of these students can be significant. You should include para-educators in meetings and conferences with special education teachers, members of the child study team, parents or guardians, and administrators.

Para-educators can be an enormous help in your classroom. Forging an effective working relationship with them can benefit you and your students.

WORKING WITH SUPPORT STAFF

Many people working in a variety of positions make up the staff of a school. In addition to teachers and administrators, you will be working with secretaries, custodians, cafeteria staff, and maintenance personnel. Each deserves your respect and consideration. Always greet the members of the support staff at your school in a friendly manner and speak and act professionally with them. Remember that they, like you, are a part of the overall team that is trying to make your school a great learning environment for its students.

It is not always easy to work together. Different personalities, different expectations, and different goals can undermine the best efforts at cooperation. Yet, quite simply, schools in which staff members work together effectively to accomplish shared goals are usually more successful than those in which staff members follow their own inclinations and move in conflicting directions.

Committees and Teamwork

When people work together to solve problems, they often accomplish more than if each individual attempted to solve the problem alone. The combined skills of the members of any group can compensate for individual weaknesses, making the group more productive than any of the individuals who compose it.

In schools, committees are regularly formed to address issues that affect policies, procedures, and programs. Whether it be a committee to revise the math curriculum of your high school, amend the discipline policies of your school, or select a new math textbook, every committee has one or more goals that its members try to attain. The degree to which committee members are able to cooperate has a major impact on the degree to which they accomplish their goals.

Effective committees share several characteristics, including:

- The committee is composed of people with diverse skills. Their collective strengths compensate for individual weaknesses.
- Clear goals are identified. All committee members understand the goals and accept the purpose of the group.
- A competent leader may be appointed, or she may emerge. The leader is respected by the group's members.
- A clear direction for achieving the committee's objectives is determined.
- All members feel that they have a stake in the process.
- Each member is given specific responsibilities. They all understand and accept their individual duties.

- Members are willing to work cooperatively to attain their goals.
- Members trust and respect each other.
- Communication is encouraged. Everyone participates in discussions.
- Committee members feel comfortable exploring and sharing ideas.
- Criticism is constructive.
- Throughout the process, committee members stay focused on their objectives.

Just as effective committees exhibit several positive characteristics, ineffective groups share several negative ones, including:

- The goals of the committee are not clear, or group members feel the goals are irrelevant.
- Individuals are not dedicated to the committee's goals or purpose.
- No leader is appointed or emerges, or she is ineffective.
- Some people dominate the discussion, giving others little opportunity to offer ideas.
- Criticism may be negative, dismissive, or sarcastic.
- People are unwilling to accept their tasks.
- Members make little effort to cooperate.
- There is little trust between members.

The most effective school committees are those in which members support the committee's purpose and work together to achieve the group's goals. They decide upon a plan of action, confront and solve problems, and meet deadlines. They work toward success.

THE ROLES PEOPLE PLAY IN COMMITTEES

When committees meet for discussion, members assume various roles. The roles an individual may assume will vary, depending on his or her personality as well as the topic of discussion. Sometimes a member assumes more than one role during the same meeting; sometimes different members assume the same role. Understanding these various roles can give you useful insight into a group's dynamics and help you to make the committees on which you serve more productive.

Following are some common positive roles that people assume during committee meetings:

- The **leader** may be appointed by an administrator, chosen by the committee members, or simply emerge. A competent leader is a guide and facilitator; she keeps people on task and makes sure that the committee moves forward in its work. She recognizes the skills of the individual group members and tries to utilize those skills to achieve the committee's objectives.

- The **thinker** is a person with ideas. He often sees problems from a fresh perspective and suggests innovative solutions.

- The **cheerleader** is a positive individual. She encourages others when spirits might be low because of obstacles that hinder the committee's progress. She provides energy and spark.

- The **recorder** is a note taker who writes down ideas, strategies, and conclusions. The recorder also keeps track of the specific tasks assigned to individuals and the deadlines for their completion. When necessary, the recorder can refer to his notes to remind other members of past actions or decisions. Some recorders are designated and others may simply assume the role.

- The **clarifier** summarizes ideas, asks questions, and restates facts in an attempt to make sure everyone understands what has been said. The clarifier reduces the possibility of misunderstanding.

- The **compromiser** helps the members of the committee avoid conflict. When ideas are in sharp opposition, the compromiser tries to find a middle ground that all group members can accept.

- The **evaluator** is a methodical person who assesses ideas in a logical manner. She is objective and bases her conclusions on facts, not opinions, and is able to see the strengths and weaknesses of various proposals.

Committees in which everyone assumes a positive role are wonderful to work with. All members are dedicated to the group's purpose and steady progress is made toward reaching objectives. In some committees, however, members may assume negative roles that undermine the group's efforts. Be aware of these roles and avoid them. Common negative roles include:

- The **dominator** tries to take charge of the committee. She interrupts others, accepts no ideas but her own, and attempts to control the group. She may lobby to be chosen as the leader, or she may act as if she is the leader, even though she is not.

- The **critic** is never satisfied. Little or nothing is acceptable. He sees negatives in everything, but seldom offers alternatives. The critic can easily become a source of turmoil in a committee.

- The **clown** views the committee's purpose and work as amusing. He jokes and finds humor in all that the group does. He contributes little and frequently leads the group off topic.

- The **know-it-all** is convinced she has the answers to all of the group's problems and questions. She can be so convinced she is always right that she may belittle the ideas of others.

- The **shirker** has no interest in the committee. He is unreliable, refuses to accept responsibility, and is unlikely to complete any of the tasks to which he is assigned.

As a member of a committee in your school, you should assume a positive role, one with which you are comfortable. Anyone who is a member of a committee has a responsibility to work with others in attaining the committee's objectives. Cooperation makes it easier to achieve objectives and is an important trait of a professional.

Working Together in Sharing

You are in a unique situation if you do not need to share space or resources with your colleagues. With most schools operating on inadequate budgets, space and resources are limited and sharing is a necessity.

As a math teacher, you may find yourself sharing more space, equipment, and materials than members of other departments. Along with sharing classrooms and work and storage space in the math department, a conservative list of items you may need to share includes: overhead projectors and transparencies, digital projectors, calculators, computers, rulers, protractors, compasses, and resource books.

The following tips can make sharing easier for you and others:

- Be considerate when sharing a classroom with other teachers. Meet with the other teacher (or teachers) and discuss how you will share the room. You may need to divide board space, shelves, and space in the file cabinet, as well as set a schedule for decorating the bulletin board and classroom. You must also decide how desks and furniture will be arranged. If a teacher moves furniture for a particular activity, she should have students put the furniture back before leaving.

- Share workspace, filing cabinets, storage, equipment, materials, and resources graciously.

- Follow procedures for checking out equipment and reserving the math or computer labs. Sign up in advance. If your plans change, cancel your reservation. No one likes to be denied access for equipment or space in a lab when it is not being used.

- If equipment you are using malfunctions, report the problem to the proper person. Do not simply put the equipment back, thinking you will inform someone later. Before you do, another teacher may try to use it and his lesson may be ruined.

- If equipment in a room you are sharing breaks down, leave a note for the next person who uses the room. Be specific in describing the problem and what you did to try to fix it, including contacting the proper support person. The next teacher will appreciate being informed, especially if she intended to use the equipment. (See "Record of Used Supplies or Malfunctioning Equipment" in Section Five.)

- Always ask before borrowing something from someone. Never take an item from someone's room without checking first. Return what you borrowed promptly and in working order. Remember to thank the lender.

- Label materials with your name and the room number.

- Do not lend supplies from a room you are sharing without first checking with those who share the room with you. People are always annoyed when they plan to use thirty rulers for a lesson and discover that the rulers are missing.

- If you are using another teacher's computer, or a computer in the teacher's lounge, always log off when you are done and return to the desktop.

- When you are working in the math department office or the teacher's lounge, be careful not to spread your books and papers all around. Leave space for others.

Sharing space, equipment, and materials is inevitable in most schools. Accepting the need to share, being considerate of your colleagues, and following procedures for using equipment and materials make sharing easier for all.

Getting Along with Others in Your School

Schools are large organizations. Each day you will be working with many staff members, sharing rooms, materials, and equipment with them, eating lunch with them, and socializing with them after school. You will be interacting with diverse individuals, many of whom will have different backgrounds, cultures, and needs. Though it is natural to like some people more than others, you should strive to build and maintain positive relationships with everyone in your school. When people get along, they are more likely to cooperate on the many annoyances, problems, and unexpected events that occur on any school day.

The following suggestions can help you get along with the staff members of your school:

- Always be polite and courteous to others.
- Treat others with respect and honesty.

- Offer people a friendly hello when you enter an office or pass in the hall.
- Greet others by name, and, if the situation calls for it, by title.
- Do not speak about other staff members in a negative manner.
- Do not gossip.
- Be upbeat and enthusiastic. Enthusiasm can be infectious.
- Give your attention to people when they are speaking to you.
- Be willing to help others, even if it means going out of your way.
- Be on time for meetings.
- Remain calm and professional during any disagreements that arise.
- Try to get to know other staff members. If some people are going out to lunch during an in-service day, join them.
- Clean up after yourself. Never leave behind empty coffee cups, napkins, or trash.
- Check your e-mail daily and respond professionally. Use correct grammar and punctuation in your responses.
- Respond to phone messages promptly and courteously.
- Be willing to share supplies, materials, and equipment with others. Few math teachers have everything they need.
- Be reliable. Do what you say you will do.
- Complete paperwork in a reasonable amount of time. Meet deadlines and do not hold up others because you have slipped behind in your responsibilities.
- Accept the customs and traditions of others.
- Support the traditions of your school. For example, if on the Friday before a football game everyone is asked to wear the school's colors, be sure to do so.
- Join school-related organizations such as the parent-teacher association.
- Attend school functions, such as concerts, drama productions, and sports events.
- Volunteer to serve on committees.

Despite your efforts to build positive relationships with other staff members, there will always be some who will be difficult to get along with. These individuals could be extremely stressed because of personal problems, be very opinionated or negative, or be overly demanding of others. Whatever the reason, they tend to be unwilling to compromise on any issue.

You can get along with difficult individuals, but it will take an extra effort. Most importantly, avoid criticizing them. Accept them as they are. Do not judge them or dwell on their shortcomings. Try to engage them in honest, two-way communication. Listen to them and try to understand their needs. Try to see things from their perspective, while encouraging them to see things from yours. Remember that as staff members of a school you share a common goal of making your school a comfortable environment for both students and teachers. That is a powerful incentive for cooperating.

When you treat others with friendliness, understanding, and consideration, they are likely to respond positively. This is the starting point of a relationship built on professional and personal respect.

Evaluations for Math Teachers

Whether you are a veteran or a first-year math teacher, you will be evaluated by at least one supervisor during the year. The frequency and procedures for evaluations vary among school districts, but nontenured teachers are usually evaluated more often than tenured teachers. Depending on your school, evaluations may be conducted by your principal, assistant principal, math department chairperson, curriculum supervisor, assistant superintendent, or your superintendent, or a combination of these.

Evaluations may be formal or informal. In a formal evaluation, a supervisor observes you and your class. She will be looking at and writing notes about your classroom management, knowledge of the subject matter, delivery of instruction, interactions with students—in short, your overall performance as a mathematics teacher. Afterward you will likely meet with her and discuss what she observed to be your strengths and weaknesses, and, most important, how you might improve any weaknesses. She will also write about her observation, which you will receive in the form of a written evaluation. In an informal evaluation, a supervisor may visit your class for part of a period, or perhaps an entire period. Although she may later discuss aspects of your math lesson with you, she will not write an evaluation.

Evaluations may also be unplanned or planned. In an unplanned evaluation, a supervisor comes to your class, unannounced, and observes your lesson. He will meet with you later to discuss what he observed and write an evaluation. The advantages of an unplanned evaluation are obvious. You present the lesson you had scheduled without any additional preparation, and you do not suffer any pre-observation apprehension. But there are some disadvantages, too. The evaluator may not be aware of the abilities of your students, particularly any special-needs students, that may require you to adjust your teaching. Also, unless he is aware of the plans you have submitted to your immediate supervisor, he may walk in on a day when you are giving a test and not teaching. In a planned evaluation, the supervisor will inform you ahead of time that he wishes to observe one of your classes. He may suggest a time or class, or you may do so. It is likely that you will meet with him prior to the observation in a pre-observation conference, as well as afterward in a post-observation conference. You will receive a written evaluation. The advantage of a planned evaluation is that you meet with your supervisor ahead of time and discuss your students and the lesson you will be teaching. He will then know to look for certain things, such as specific arrangements you have made for a student with special needs. He may also offer you suggestions for the lesson, which can help you present an outstanding math activity. The disadvantage, of course, is knowing that you will be observed, which may lead to undue stress and anxiety.

Few teachers enjoy being observed. Many new teachers feel that administrators are looking for shortcomings in their teaching. Experienced teachers sometimes feel that supervisors do not offer helpful suggestions. This can be especially true of math teachers who are observed and evaluated by supervisors who lack a strong background in mathematics.

Despite the uneasy feelings that teachers may have about being evaluated, evaluations are an important part of developing and maintaining professional skills in the classroom. During an observation, your observer will become acquainted with your lesson planning, instructional delivery methods, teaching style, management of your classroom, and your interaction with students. By identifying both your strong points and areas that might need strengthening, an evaluator can help you to improve your classroom management and teaching skills. Evaluation thus becomes a means for professional growth.

HOW NOT TO BE NERVOUS DURING AN OBSERVATION

It is normal to be somewhat nervous before an upcoming observation. Even veteran teachers may feel some apprehension. After all, classroom dynamics can change when someone else comes into the room, students may misbehave, and equipment can malfunction. Anticipating the unexpected results in worry.

Doing the following can help you reduce nervousness before an observation:

- Design an interesting lesson with clear objectives that engages students in math.
- Make certain that you have all the materials and equipment you will need.
- Check that equipment is working and that you know how to use it.
- Follow procedures for your observation. Hand in necessary paperwork on time.
- Be ready for the unexpected and react in a professional manner. Evaluators are looking at your overall performance in your duties as a teacher, and not just one or two areas. To an observer, a difficult student's sudden outburst is not as important as your reaction and handling of the situation.
- Be confident in your preparation and knowledge.

Knowing that you are ready for an observation eases worry and allows you to focus on your teaching. Remember that the goal of the observation is to help you become a better teacher.

PREPARING FOR AN OBSERVATION

Knowing when you will be observed affords you the opportunity to prepare. When your supervisor contacts you about an upcoming observation, if possible, find out the day and time she plans to observe you. If you have a choice, schedule the observation for a time when you can provide a great lesson with a hard-working class. Try to develop a lesson in your current unit of study that includes components that are essential to outstanding math classes. Some examples include a lesson in which students use graphing calculators to explore and organize data; use pattern blocks to discover geometric concepts; or use computer software (for instance, Geometer's Sketchpad from Key Curriculum Press, www.keypress.com) to investigate the properties of parallel lines, transversals, and their angles. Any lesson that encourages students to become active participants in learning will result in an interesting class.

If technology is a major component of your lesson—maybe students need to use the Internet to complete an investigation of fractals—have a backup plan in case the technology malfunctions or is inaccessible. Perhaps you intend to use a computer and whiteboard to demonstrate virtual manipulatives. If the equipment is not working, be ready to use an overhead projector with the appropriate manipulatives to complete your lesson. Your evaluator will understand that malfunctioning equipment is beyond your control and she will be impressed by your smooth implementation of another method of instruction.

Once you have decided on a lesson for your observation, write your lesson plan using the standard lesson-plan format for your school. If your school does not have a format for lesson plans, see "Daily Lesson Plans" in Section Ten for a model. Be sure to include necessary information such as your name, date, course, time, objectives, standards, materials and equipment, procedures, assessment, and any homework assignments. Provide a copy of your plan to your supervisor before the observation.

AN EVALUATION CHECKLIST

Most supervisors focus on specific criteria when they observe teachers. Being aware of the areas on which your supervisor will concentrate will help you prepare for the observation.

Though evaluative criteria vary among supervisors, most assess math teachers' expertise in the following areas:

- Lesson preparation
  - Demonstrates knowledge of mathematics
  - Demonstrates knowledge of district curriculum

- Demonstrates knowledge of math standards
- Selects appropriate goals for class
- Selects appropriate content for lessons
- Designs interesting and motivating lessons
- Utilizes available instructional materials and equipment

○ Lesson presentation

- Makes purpose of the lesson clear to students
- Reflects the lesson plan
- Uses appropriate instructional techniques
- Uses proper grammar and appropriate mathematical terminology
- Delivers effective instruction of mathematical concepts and skills
- Incorporates technology
- Makes full use of class time
- Keeps students on task
- Encourages students to be active participants in activities
- Maintains student interest
- Demonstrates effective questioning techniques
- Asks relevant questions
- Provides reinforcement
- Includes critical thinking activities
- Uses various strategies in working with students
- Has necessary materials and equipment available
- Summarizes lesson and provides closure
- Assesses student learning
- Indicates homework and follow-up activities

○ Student-teacher relationships

- Works well with individuals and groups
- Motivates students to succeed
- Is sensitive to individual student needs
- Uses positive, supportive statements with students
- Demonstrates rapport with students
- Provides opportunities for all students to be successful

○ Classroom environment

- Manages classroom routines effectively
- Maintains an orderly classroom

- Maintains proper discipline
- Maintains a safe, comfortable environment for students
- Maintains a neat, attractive classroom
- Uses bulletin boards and visual aids to enhance learning

THE PRE-OBSERVATION CONFERENCE

During a pre-observation conference, you meet with your supervisor to discuss the lesson she will observe. Bring a copy of your lesson plan, any activity or homework sheets the students will be working on, and any notes about the class you would like to share. Also be sure to bring any items or materials your supervisor requests, such as your grade book.

In addition to talking about your lesson, describe your class and any situations of which your supervisor should be aware. If, for example, the class contains several basic skills students, mention this and any special strategies you have used to meet their needs while continuing to meet the needs of the other students. Be personable and willing to consider any advice your supervisor may offer. If she identifies specific criteria that she intends to focus on during the observation, be sure to note and address them during the lesson.

The pre-observation conference provides an opportunity for you and your supervisor to share concerns and expectations about your upcoming observation. It is an important meeting that can help lay the foundation for an excellent evaluation.

THE OBSERVATION

On the day of your observation, dress professionally. When your supervisor arrives at your classroom, greet her and direct her to an empty seat. Provide her with a text and any materials that your students will be using during the math lesson. If your students ask why your supervisor is in the class, simply say "Ms. Jones is here to observe our class learn math today."

Supervisors have different styles for observations. Some write notes throughout the period; others write few notes. Some remain sitting in the back; others will circulate around the room and talk with students, especially if students are working in groups.

Present and teach your lesson as if the supervisor was not there. As you become involved in teaching, any nervousness you have will ease and you will soon be interacting with your students. Should a problem occur, address it promptly and professionally, and return to your lesson. Supervisors are well aware of the things that can happen in a classroom. Remember that your supervisor is looking for your strengths as a math teacher and ways to help you improve any weaknesses.

THE POST-OBSERVATION CONFERENCE

At the post-observation conference, you will discuss with your supervisor what she observed during your math lesson. She may note things, both positive and negative, of which you may not have even been aware. Accept credit for the good points, and accept the challenge for improving areas in which you can do better. Try to resolve any areas of dispute, but keep in mind that the recommendations your supervisor makes are in the best interests of you and your students. Always be professional during discussions; arguing seldom solves differences of opinion.

After your post-observation conference you will receive a written evaluation of your lesson. Read the evaluation carefully. If you feel that it is inaccurate, share your feelings with a respected colleague, preferably another teacher in the math department who is likely to understand any references to math. Most schools have a procedure for refuting points made on evaluations. If you believe that you must respond to your evaluation, be sure to follow the proper course.

Always keep a copy of your evaluations and follow the recommendations of your supervisors. Use their suggestions as a means to improve your teaching.

THE END-OF-THE-YEAR EVALUATION

The purpose of an end-of-the-year evaluation is to assess your overall effectiveness as a member of the staff of your school. This evaluation is based on your work throughout the year. In addition to your teaching, your end-of-the-year evaluation should include any committees on which you served, courses or conferences you attended, and extracurricular activities you directed or assisted with. It may also include comments about your ability to work with other staff members and your contributions to the school community.

An end-of-the-year evaluation may also offer suggestions for the next year. Try to implement any recommendations and work towards further developing your professional expertise.

Evaluations provide a means of identifying strong and weak areas of your teaching. They enable you to improve your performance in the classroom and make you a more valuable member of your staff. Acting positively on the recommendations of your supervisors will help you become a better math teacher.

Becoming a Mentor for New Math Teachers

Mentors are valued members of the staffs at their schools because they nurture new teachers. The guidance and support of a mentor can help a new teacher meet the challenges of his first year in the classroom and go on to build a successful career.

Districts vary in their recruitment of mentors. Some have an established mentoring program, in which administrators approach veteran teachers who are respected for their professionalism and request that they mentor new teachers. Such districts usually have qualifications for mentors and expectations for the mentoring program. Other districts have informal programs in which veteran teachers simply assume the role of an unofficial mentor for new teachers. Whether district sponsored, unofficial, or somewhere in between, mentoring is vital for helping new teachers develop the skills they need.

Although education programs in college and the experience of student teaching are designed to prepare novice teachers for the classroom, the demands and responsibilities facing first-year teachers can be overwhelming. New math teachers, like all new teachers in a school, need to learn the policies and procedures of their school, how to manage a classroom, and how to apply the methods and teaching techniques they learned in their college courses. But new math teachers must also learn how to make mathematics—which can be abstract for many students—an interesting and relevant subject. They need to learn how to design lessons that satisfy math standards, utilize models and manipulatives in instruction, incorporate technology into their lessons, and encourage students to communicate mathematical ideas. They need to learn how to thrive in a math classroom, as well as how to become valued members of their staffs and school communities.

Veteran math teachers can help new math teachers by being role models and sharing their professional expertise. They can provide the advice and guidance a new teacher needs to begin a successful career.

RESPONSIBILITIES OF A MENTOR

Mentors have numerous and varied responsibilities. If you are a mentor, you should:

- Attend training sessions to hone mentoring skills. These sessions offer information, ideas, and activities to guide you through the mentoring process.

- Serve as a role model both in the classroom and as a professional in your school community.

- Meet with your mentee before school begins, then a few times each week at the beginning of the school year, and weekly or as needed as the year progresses.

- Establish a relationship that encourages your mentee to ask questions without fear of ridicule, even if the answers to the questions seem obvious to someone with teaching experience.

- Acquaint your mentee with your school, including the math curriculum, math standards, school and district policies and procedures, evaluation process, school calendar and events, and testing procedure.

- Offer assistance and emotional support.
- Visit the new teacher's classroom. Offer suggestions on classroom activities, instructional techniques, pacing, and classroom management.
- Allow your mentee to observe your classroom to see how you teach and manage routines.
- Make available the math materials and resources that will help your mentee develop effective teaching strategies and techniques.
- Encourage your mentee to keep a record or list of needs, questions, and concerns that she can then discuss with you.
- Encourage your mentee to become involved with professional development opportunities in your district and also to join professional organizations for mathematics teachers.

Many people who are unfamiliar with the mentoring process mistakenly believe that new teachers gain most, if not all, of the benefits. Mentors, however, benefit as well. In order to help novice math teachers, mentors frequently explore new instructional strategies, ideas, and technologies. They often reexamine their own methods and try new approaches. The mentors grow as professionals and emerge as leaders on their staffs.

RESPONSIBILITIES OF A MENTEE

Just as a mentor has numerous and varied responsibilities in the mentoring process, so does the mentee. If you are a mentee, you should:

- Meet with your mentor regularly and establish a relationship built on honesty and trust.
- View your mentor as a confidant and source of support.
- Ask questions. No one can answer a question that you do not ask.
- Keep a notebook to record areas of concern and questions you may have. Refer to the notebook during discussions with your mentor.
- Observe your mentor (and other veteran math teachers, if possible), and allow your mentor to observe your classroom.
- Consider new ideas, and accept those that you feel you can use to improve your teaching.
- Develop the skills and techniques necessary to enhance student learning.
- Set realistic expectations for yourself, your mentor, and your students.
- Continue to grow and embrace the profession of teaching.

The novice math teacher has much to gain from a mentor. He learns from an expert, and becomes knowledgeable in techniques for classroom management, lesson development, and teaching methods. He becomes aware of his curriculum and math standards and how to implement the standards in his lessons. He becomes familiar with the various resources and technology a math teacher utilizes to provide effective instruction. Most important, the novice math teacher grows professionally, acquiring the skills and attitudes that will help him to one day become the expert. (See "Especially for the First-Year Math Teacher" in Section Two.)

The benefits of mentoring extend beyond the novice teacher and his mentor. Because of the support of his mentor, a new teacher will be more effective in the classroom, which in turn results in a more productive learning environment for students. The school benefits, too, as the new teacher acclimates to the routines of the school more quickly. Without question, mentors are a valuable resource in any school. Consider becoming one!

After School and Beyond

A teacher's day does not begin and end in the classroom. Students may need help after school, extracurricular activities and clubs need advisors, athletic teams need coaches, and Friday night dances need chaperones. It is likely that you will be involved with some of these and similar after-school activities.

Avoid the mistake of looking upon such activities as simply adding to the length of your day, and instead view them as opportunities to help students and get to know them better outside the classroom setting. In many activities and sporting events, you will also meet and work with parents, guardians, and staff members you may not normally see. Teachers who help students after school and spend time with extracurricular activities are invaluable to any school. (See "Expanding Your Role as a Math Teacher" in Section Fourteen.)

PROVIDING AFTER-SCHOOL MATH HELP

There will be times throughout the year that some of your students will need extra help to master math concepts and skills, or they will need to make up missed tests or quizzes because of absences. Unless you have a free period when you can meet with these students, you will probably meet with them after school.

Following are suggestions for providing after-school help:

- ◎ Always meet in the same room. This eliminates the problem of a student having to find you, and also prevents her saying, "I looked for you and you weren't there."

- If you must attend an unexpected meeting, take a phone call, or sit in on a conference and cannot meet with students after school, leave a note on the outside of the door explaining that you cannot meet today. State the next time that you will be available after school.

- Establish a rule that you will wait ten minutes for students to report for extra help. Do not waste time waiting for students who say they are coming but do not show up.

- It is best to share a room with another teacher when you work with students after school. Having two teachers in the room alleviates the problem of your being the target of a story that is completely contrived.

- If you must be in a room by yourself, meet with a few students at a time. Sit in the front of the room in full view of the door. Leave the door open.

- Ask students to check with you before coming for extra help. This reduces the problem of students feeling that they can come for help whenever they please.

- Check with your principal about the use of cell phones after school. If it is permissible, allow students to call their parents or guardians to inform them that they are working with you after school (and to arrange for transportation home if necessary).

- Keep a log of students who report for extra help. Include the date, time they reported, and the time they left. Some students may tell their parents that they are staying after school for extra help when they are in fact at the mall.

- Avoid too much socializing with students who come for after-school help. Some students may enjoy your company and want to spend time with you even though the time spent is after school. Others may be going home to empty houses and prefer to stay in school with you. Though after-school help will be conducted in a relaxed atmosphere, keep the session focused on math help.

- Politely discourage students who do not need help from visiting after school. They may be waiting for a friend who is involved in an activity, or they may simply have time to pass. The purpose of after-school help is schoolwork.

- Avoid permitting students to take advantage of your willingness to help them. Do not allow them to rely on your assistance rather than relying on themselves to do the hard work necessary for learning math.

- If a student cannot come after school for help because of another activity or a part-time job, you may wish to meet with him before school, but only if it is convenient for you. Mornings can be hectic, especially if you have a long commute or lunches to make and your own kids to send off. Perhaps you can meet with this student at some point during the school day.

The time you spend providing extra math help can be very productive. For some students it can be the difference between mediocre and strong progress in math.

EXTRACURRICULAR ACTIVITIES

Schools always need teachers to coach athletic teams, advise clubs, and chaperone special events. Teachers who are willing to give their time to such extras earn the appreciation of administrators, colleagues, students, and parents and guardians.

If you enjoy and have had training in a particular sport that is offered at your school or district, you may consider coaching. Unless you have coached before, however, you should first seek a position as an assistant coach. This will give you the chance to gain experience coaching at your school. Be sure to familiarize yourself with any manuals or guidelines for coaches your school may have.

If you enjoy a particular activity—for example, chess, math trivia, or computer graphics—you may want to supervise an after-school club for students. As with coaching, be sure to familiarize yourself with any guidelines your school has for club advisors. Furthermore, if you are advising a club in which athletics are involved, such as an after-school dance club, check with your principal and school nurse regarding your qualifications, required physicals for students, or health forms that parents or guardians must complete.

If you are a chaperone for a special event, perhaps a school dance, drama production, or science fair, ask the individual in charge what your duties are. Make certain you are at your assigned place at the proper time. Remember, others are depending on you to help make the event a success.

Teachers who regularly assist in extracurricular activities for students become important members of their school community. They are valued members of their staffs, not just because of their responsibilities in the classroom but because of their overall dedication to their students.

Quick Review for Becoming a Valued Member of the Staff

In just about every school, people know who the best teachers are. These teachers are often described as being dedicated to teaching, committed to students, and willing to work with others to meet the daily challenges of education.

The following suggestions highlight how you can become one of these teachers:

- Cooperate with administrators, colleagues, and support staff to find solutions to problems and ensure that your school provides quality education to all its students.
- Understand and follow the chain of command in your school.

- Work effectively with other teachers and staff members.
  - If you are a member of a team, support your team's goals, rules, and procedures.
  - If another teacher works with you in your classroom, decide together how you will share responsibilities and manage the class.
  - If a para-educator works in your class, try to involve her in the class as much as possible.
- Strive to assume positive roles on teams or committees and avoid negative roles.
- Be willing to share work and storage space, materials, supplies, and equipment with others.
- If you share a classroom with other teachers, exit the room on time and leave it clean and orderly. No one likes to be kept waiting at the door or have to pick up another class's mess.
- Work to get along with others in your school, even those individuals deemed hard to get along with.
- View evaluations as guidelines for improvement.
- Become a mentor to share your professional expertise and experience with new math teachers.
- Do the extras.
  - Offer after-school math help to your students.
  - Be an advisor to clubs and activities.
  - Be a coach.
  - Be a chaperone at school events.

The best math teachers do not only teach math well, they also work cooperatively with others to ensure a successful educational environment for students. By devoting themselves to improve their school and school community, they truly become valued members of their staffs.

SECTION FIVE

Organizing for Success

*L*ike almost every other math teacher, you probably feel there is not enough time in the school day to accomplish all you need to do. Between planning interesting math lessons, providing quality instruction for students of diverse abilities and backgrounds, managing your classes, and interacting with students and colleagues, not much time is left for the many other tasks that are required of you—marking papers, recording grades, attending meetings, serving on committees, ordering supplies, maintaining calculators and computers, checking e-mail and voice mail, keeping parents and guardians informed of their children's progress, and on and on.

Because time is always short, it must be managed efficiently. You must organize each day for success. Establishing practical routines, prioritizing tasks, and handling daily responsibilities successfully will help you to achieve all that must be done.

Your Master Schedule for Organization

Your school day is divided between teaching and nonteaching responsibilities. As much of your day at school centers around your time in the classroom, you must use other parts of your day to manage your nonteaching duties. Before and after school, planning periods, free periods, and lunch provide a significant amount of time, which, if used effectively, will enable you to attend to the many nonteaching tasks that fill a teacher's day.

To take charge of your time, obtain a calendar or planner. These come in many forms and combinations, and may be paper or electronic. Select the kind you feel most comfortable using. Your calendar or planner will be the place where you record dates, events, and tasks you must complete—anything that will affect your teaching schedule.

Most school districts provide calendars at the beginning of the year. Use your district's calendar as the basis for your own, which should focus on information pertinent to you and your classes. You may find it helpful to highlight major events. For example, you could use red to mark important dates such as back-to-school night. If you are using a paper calendar, you can attach small adhesive notes to add information about a specific date. Checking your calendar every morning and a few times throughout the day will keep you aware of things that must be done.

Dates and events you record on your calendar should include the following:

- Days when school is not in session
- Half days
- Faculty meetings
- Math department meetings
- Back-to-school night
- Due dates for units or daily lessons plans
- Professional development days
- Beginning and ending dates of marking periods
- Progress report dates
- Report card dates
- Deadlines for midterm notices for parents and guardians
- Standardized testing dates
- Parent-teacher conferences

Update your calendar as you learn of new events or special dates. Throughout the year you may find it necessary to include the following:

- Workshops, seminars, and conferences
- Meetings with parents or guardians or administrators
- Assemblies
- Trips
- Special events, such as math contests, rehearsals for concerts or plays, or multicultural celebrations
- Special days at your school, such as "Spirit Day"
- After-school activities or clubs you advise
- School dances at which you will chaperone
- Personal days

○ Personal commitments such as a doctor's appointment, your child's awards assembly at her school, or a closing on a new house (Note: Do not include personal commitments of a sensitive nature if others might see your calendar.)

Maintaining a current calendar is essential for planning. An accurate calendar of events can keep you informed of special activities in your school and help you to avoid scheduling conflicts—for example, giving a test on a day after several of your students performed in a band concert the evening before, or committing to an out-of-district math conference on a day of standardized testing at your school. An accurate calendar also helps you to meet deadlines and eliminates the problem of having to sort through piles of paper to find the date and time of a seminar you are to attend.

Think of your calendar as a broad time line of special dates, events, and activities. It lays out the big picture, making it easy for you to fill in the details.

The Value of an All-Purpose Binder

Many math teachers, particularly those who teach in different classrooms, utilize all-purpose binders to help them stay organized. A large three-ring binder with pockets and dividers is a good choice. In the pockets you can store pens, pencils, a calculator, CDs, flash drives, a protractor, a compass, notepads, and adhesive notes. Within the sections you can store daily lesson plans, worksheets for photocopying, memos, address lists, and copies of tests, quizzes, and answer keys. You may divide the binder's sections according to class, topic, or unit, and may even be able to store your calendar or planner in your binder.

Taking your binder to your classes helps to ensure that you will have the materials you need for teaching. Binders are an excellent tool for organization.

Practical Routines

The establishment of practical routines is essential to organization. Routines provide a reliable method for accomplishing the daily tasks that make up a large part of each teacher's day and can save considerable time.

You should set up practical routines for your classroom as soon as possible. Routines might include the following:

○ Entering the classroom and starting a do-now

○ Taking attendance while students are working on the do-now

- Distributing and collecting forms
- Distributing and collecting assignments
- Distributing and collecting materials and supplies
- Obtaining and returning calculators
- Logging on and off computers
- Procedures for moving furniture in class to form groups
- Policies for leaving class, including the use of hall passes
- Leaving class at the end of the period

Practical classroom routines minimize confusion and promote efficiency. They help your days run smoothly. (See "Establishing Efficient Classroom Routines" in Section Seven.)

You should also establish practical routines for the rest of your day. Begin before you leave for school in the morning by thinking ahead to your upcoming day and making certain you have all of the materials you will need. Do a mental check—briefcase or tote, binder, laptop, papers, lunch, that resource book you had at home but need for class. Stepping into your classroom and realizing you left a set of transparencies at home that you need for your first period class will surely undermine the positive start you had planned for the day.

Upon arriving at school, check your mailbox in the office, your e-mail, and voice mail. Note any special announcements of dates or events. Record them on your calendar, including the time. Also note any messages or correspondence to which you must respond. If you have time, answer immediately. Otherwise, prioritize them in order of importance and reply later. You might set aside time during your free period each day to respond to high-priority messages or requests, and respond to others within a day or two. You might also reply to messages after school. Always respond to messages as soon as possible, though. No one likes to wait overly long for a response.

You may prefer to use the following reproducible, "Daily Reminders," to keep track of your tasks. Simply write down the things you must do on the sheet, and place the sheet in your binder. Refer to it later in the day when time becomes available. The reminders can be especially helpful on busy days.

Daily Reminders

Date _____

Phone Calls:
E-Mails to Send:
Memos to Write:
Papers to Grade:
Grades to Record:
Activities to Plan:

Daily Reminders (continued)

Tests to Create:
Copies to Make:
Conferences:
Meetings:
Materials to Obtain or Order:
Other:

If your schedule permits, you should also take a few minutes before students arrive to set up your classroom for the day's activities. (This may not be possible if you teach in different classrooms throughout the day.) Writing the do-now and day's objectives for math on the board, reviewing lesson plans, and double-checking that the equipment you will need for the day's lesson is available and operational will help to assure a smooth opening to class.

Setting up a realistic schedule to manage your nonteaching tasks will bolster efficiency. Use your planning period for planning, but set aside other open slots of time for such duties as responding to messages, grading papers, recording grades, ordering supplies, filing papers, making photocopies, and updating your assignments on your school's Web site.

Make efficiency your goal. Use your nonteaching time in school to complete school-related tasks.

HOW TO AVOID LETTING THE "LITTLE THINGS" PILE UP

Sometimes we can become so concerned with managing important tasks and problems that we let minor ones accumulate. If neglected too long, a host of minor tasks may require a substantial amount of your time. The following tips can help you keep the "little things" from piling up:

- ◎ Recognize the difference between major and minor tasks. A minor task might be correcting the makeup quiz of a student who was absent, whereas a major task might be writing a daily lesson plan. Do not try to address a major task in a few minutes. You are unlikely to handle it effectively and will only waste time.

- ◎ Set aside some time each day to address minor tasks. Try to attend to such tasks as they come up.

- ◎ If you are responding to a question via e-mail, make sure you have the answer before responding. Starting to write a response and then having to research the answer can quickly exceed the bounds of a minor task.

- ◎ Gather all the information you will need before attempting to solve a minor problem. For example, if you call a parent to respond to her question about her son's grade on his last math test, have the test on hand to refer to.

View routines as organizational aids. Practical routines enable you to be more productive. Your time will not only be well spent but well managed.

Organizing Your Classroom

An organized classroom facilitates learning. In organizing your classroom, you should eliminate any conditions that might undermine your students' achievement.

Start by arranging furniture so that all students can clearly see the board, screens, and you as you teach. Learning stations, computer tables, or displays should be set at the sides or in corners of the room, out of the way of classroom traffic and out of the direct line of sight of students at their desks. Placing computers at the front of the room can be a distraction. If a student is using a computer to view a geometric principle, other students may be watching what is on the screen rather than focusing on their own work. Furniture should be arranged in a manner that allows for easy access around the room. (See "Arranging Furniture to Enhance Math Learning" in Section Two.)

Your desk should provide you with a practical workspace. It should have a few drawers, and be big enough for your laptop and books, as well as provide you with an adequate area to work. Trays for homework, papers to be graded, and papers to be photocopied can keep your desk free of clutter.

Bookshelves, file cabinets, and storage cabinets are essential for storing materials and supplies. If you feel your room lacks space for storage, contact your supervisor. Perhaps a spare cabinet is locked away in a stock room, or another teacher has a bookshelf she no longer needs. Without sufficient storage space, materials will likely wind up in boxes at the back of the room where they might be forgotten or, in time, disappear.

Store similar materials and supplies together. Paper, for example, should be kept in one area. Protractors, rulers, and compasses should also be stored together, as should computer software and printer cartridges. Teacher resource books should be on a shelf near your desk, whereas general reference books for students (consider yourself a fortunate math teacher if you have these) should be placed on shelves from which students can access them.

Materials, supplies, and manipulatives should be stored according to the frequency of their use. Seldom-used items should be placed at the rear of a storage cabinet and items used often should be kept in the front where they can be easily reached. You must also consider security. Inexpensive items, such as protractors and compasses, may be stored in boxes on shelves; however, calculators, batteries, and other electronics should be kept in locked cabinets.

You will also need storage space for equipment. All equipment should be secured but accessible. If you have your own overhead projector, for example, try to store it on a cart in a front corner of the room. Leaving equipment in the back risks tampering or vandalism.

When everything has a place, and is stored in that place when not in use, your classroom will be neat, uncluttered, and efficient. Encourage your students to take pride in the classroom and help keep it organized. Guide them in the establishment and maintenance of practical routines. Remind them to always return materials to their proper places. Instruct them to place bags, knapsacks, and other materials under their chairs so that they do not clutter aisles. An organized classroom is a more pleasing and effective environment for learning than one in which sound routines are not maintained, and where materials, equipment, and students' belongings are scattered about.

Special Organizational Considerations When Sharing a Classroom

Organizing a classroom of your own is challenging. Organizing a classroom that you share with one, two, or more teachers is even more challenging because you must balance your organizational plans with theirs. Cooperation and practicality are essential.

When you share a room, you share everything in the room, including the teacher's desk, storage space, materials, supplies, and equipment. You must divide the use of board space, bookshelves, storage units, and bulletin boards. There are many decisions to make.

Before anyone begins organizing a shared room, everyone who shares it should meet. General guidelines for using the room should be discussed and agreed on, including:

- Each teacher should obtain her own key for the classroom, as well as keys for storage cabinets.

- Students should not sit at the teacher's desk, nor look for items in the drawers. As the desk is shared by other teachers, some items in or on the desk may not be meant for students to see.

- The arrangement of furniture should support the teaching style and methods of all teachers who use the room. If a teacher moves furniture during her class, she should return it to its original place before leaving.

- Basic supplies and materials should be pooled.

- Common items such as staplers and hole punches should be labeled with the room number. Should they be inadvertently taken from the room, labeling makes their return more likely.

- All materials and equipment should be stored in specific places and returned to their storage place at the end of class.

- Common materials and supplies should not be taken to other classes. If a teacher loans equipment or supplies to someone else, she should inform those who share the room.

- Teachers should erase chalkboards or whiteboards before leaving class.

◎ If a teacher uses the computer, he should log off and return the screen to the desktop before leaving class.

◎ Teachers should leave the room clean and orderly.

◎ When teachers leave the room, they should take everything they need with them. They should not interrupt another teacher's lesson because of a forgotten coffee cup.

◎ If you find that you have too many materials to carry to your classes, consider using a wheeled cart. You could place everything you need for the entire day on the cart and push it with you as you travel from class to class.

To help you manage supplies and equipment of a shared room, use the following reproducible, "Record of Used Supplies or Malfunctioning Equipment." Keep a copy of this sheet in an agreed-on place—perhaps a drawer in the teacher's desk or taped on a corner of the desk—and use it to record depleted supplies or equipment that needs repair. For example, if a student used the last tissue in the next-to-the-last box during Period 4, the teacher would fill in the date and period, note that she took the last box of tissues from the storage cabinet, and that she will order additional boxes. She would include her initials. Another example is a student who was unable to access the Internet from a computer in class. The teacher would fill in the date, period, a description of the problem, and how she tried to resolve it. After being unable to access the Internet herself, she would contact the tech person, telling him of the problem. Recording the information on the sheet informs the other teachers about the computer problem and saves them from wasting their time trying to rectify it.

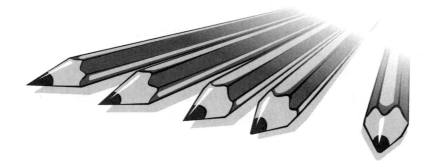

Record of Used Supplies or Malfunctioning Equipment

Room # _____

Date	Period	Problem	Resolution	Initials

Though sharing a classroom can be challenging, it can also result in a positive teaching experience. When two or more people cooperate in organizing a room, the workload is divided and the benefits are shared by all.

An organized classroom supports teaching and learning. Organization has a positive effect on you and your students, and is a major factor in the efficient operation and success of a math class.

Organizing Files

Each day you will handle a variety of memos, announcements, letters, and forms. Some will be paper; others will be electronic. All will need to be handled, saved, or discarded. You must develop an effective filing system for both paper and electronic information.

YOUR FILE CABINET

You need a file cabinet. If you must share one, be sure that at least one drawer is yours. Everyone who shares the cabinet should have his or her own key, and it should be understood that the cabinet must be locked at all times. A locked cabinet safeguards confidential information.

An organized file cabinet is a fine managerial tool. Obtain several folders and arrange them in a manner that best suits your needs. You may have a separate file for each of the following:

- Student information, including addresses, contact information, book numbers, calculator numbers, and specific student records such as IEPs, 504 plans, and medical information
- Attendance records
- Progress reports
- Report card grades
- Seating charts
- Emergency substitute plans
- Standardized test information, including student scores
- Memos and correspondence to which you will need to refer
- Correspondence with parents and guardians
- General forms, including hall passes, detention forms, and sign-out sheet
- Student and teacher handbooks
- Budget information
- State standards and district math goals

- Curriculum guides
- Your evaluations
- Records of your professional development
- Unit plans according to subject (Maintain one folder for each unit. Each folder should include master copies of daily lesson plans, tests, quizzes, answer keys, activities, worksheets, projects, and rubrics. The advantage of this, of course, is that all materials for a specific unit are contained in one place.)

When organizing your files, be sure to label each one accurately and clearly. Consider using color-coded labels. All algebra materials might be blue, geometry materials might be red, and calculus materials might be yellow.

Avoid placing so much material in your file cabinet that the files become unwieldy. Go through your folders and files periodically and discard materials that are no longer necessary. You may, for instance, take home the files of units you have already taught and store them there. A good practice is to go through and discard unnecessary materials at the end of the year.

ELECTRONIC FILES

Along with managing traditional paper files, you will need to manage electronic files on your computer. You must be able to access, save, share, and delete files efficiently.

The following tips can help you handle electronic files:

- Set up your user name and password in order to log on to your district's computer network and manage electronic files at school. You may also be able to access your school files from home.
- Make sure your e-mail account at school is fully operational, and that you understand its capabilities.
  - If your e-mail program does not already have folders for different e-mail activities, including an inbox, sent items, and deleted items, set up folders yourself. Consider creating an archive where you can place messages you need to keep. You may also create folders for different types of e-mail you receive, for example, a separate folder for e-mails from parents and guardians and another for e-mails from staff members.
  - Set up filters so that incoming e-mail is automatically placed in the appropriate folder. (If you do this, remember that you will need to check for new messages in each folder.)
  - Check your e-mail regularly, at least a few times each day.

- Try to respond to messages promptly. If you are unable to respond right away, leave the message in your inbox and respond as soon as you can.

- Create an address book and keep it updated.

- When writing e-mail, use the subject line with an accurate term or phrase and keep your messages concise.

- After answering a message, decide if you need to save it. Otherwise delete it.

- Periodically go through your e-mail folders and delete items that are no longer necessary. Unnecessary messages quickly accumulate and take up memory and hard disk space, which can hamper the efficiency of your system.

○ Create folders for data stored on your computer. Creating all folders in one place, for example, in My Documents, makes it easier to find folders and make backup copies.

- Store all files related to a specific topic in the same folder.

- Use clear, logical folder and file names. For example, the name of the folder for the first unit of your pre-algebra class might be Pre-Algebra Unit 1, Operations with Integers.

- Always create backup copies of data on a removable storage device, such as a flash drive or CD. If your hard drive crashes or becomes infected with a virus, you will be thrilled that you have backup files.

- Maintain up-to-date virus protection and Internet security for your computer.

- Periodically go through your files and delete those that are no longer necessary.

One day, computer storage systems may make traditional file cabinets obsolete. That day is not here yet, and maintaining both paper and electronic files is necessary if you are to manage information effectively.

Necessary Information to Maintain

You can easily become overwhelmed by the amount of information you must manage throughout a school year. Trying to save copies of everything quickly leads to folders, both paper and electronic, becoming so full that finding needed data is difficult and time-consuming. To avoid an overload of stored information, you need to save only important data and discard the rest.

Keep the following information:

- State standards and district math goals
- Curriculums
- Master copies of units, including lesson plans, student worksheets, activities, projects, tests, quizzes, rubrics, and answer keys
- Special lesson plans for math events
- Math exams
- Handbooks and manuals
- Plans for first-day activities
- Attendance records
- Contact information
- Student locker numbers, book numbers, calculator numbers, and so on
- Copies of progress reports and grades
- Standardized test scores
- Copies of important materials sent to parents and guardians
- Copies of important correspondence with administrators, colleagues, and parents and guardians
- Your teacher evaluations
- Professional development information
- Budget information and copies of orders for materials and supplies
- Announcements of important dates or events
- Old worksheets that may be a source of future ideas
- Samples of student work
- Materials from in-services and workshops

Discard general information and information that is no longer relevant, such as the following:

- Announcements and reminders that do not specifically affect you or your classes—for example, an announcement about an upcoming history trip that is being handled by a history teacher. You will want to log the date on your calendar, however.
- Items that are no longer relevant, such as the lunch menu for last month or a flyer for last week's flower sale.
- Items no longer relevant once the school year ends, including calendars, student handbooks, your schedule, seating charts, general student information, and directives.

Discarding irrelevant materials substantially decreases the volume of your stored information. Trying to keep copies of everything that is placed in your mailbox or arrives in your e-mail will only burden you with having to handle too much information.

Managing Your Paper Load

Most teachers find their daily load of paperwork to be one of their most demanding responsibilities. Tasks such as writing lesson plans, grading papers, keeping records, and managing the seemingly endless forms, correspondence, and memos consume a major portion of each day. Effective handling of paperwork eases the burden and enables you to spend more time directly related to teaching.

Following are some suggestions for managing your paper load:

- Attend to announcements, memos, and forms as soon as you receive them.
- Establish routines for distributing and collecting papers. Consider having reliable students in each class help you.
- Have separate trays on your desk for homework, classwork, and forms.
- Try to handle each set of papers once.
- Grade papers as soon as possible. Do not allow stacks of papers to pile up.
- Grade the papers of a class in one sitting. Record the scores as soon as you are done with grading.
- Return papers to your students promptly after grading.
- File forms and information you must keep when you receive them or as soon as possible.
- Organize important papers in a binder.
- Keep the papers of different classes separated.
- Keep answer keys on hand for each worksheet, test, or quiz.
- Have available additional copies of worksheets and activities for students who may misplace theirs.
- Try to stagger tests, projects, and major activities so that you are not overwhelmed with too many papers to correct at the same time.
- Encourage students to make up work for absences promptly.
- Take the papers of one or two classes home each night. Trying to do too many will only frustrate you.
- Use free periods in school to correct papers and record grades.

Managing your paper load should be one of your daily routines. It is an important part of being organized.

Taking Control of Time

There is never enough time in the school day. As you cannot create more time in your day, you must use the time you have to get as much done as possible.

The following tips can help you take control of time:

- List and prioritize the tasks you do each day.

- Arrange tasks to fit your schedule. For example, if you are a "morning" person, do your planning before students arrive and classes begin. If you prefer to plan at home, utilize your time in school to manage other tasks.

- Use brief chunks of time effectively. In just a few minutes you can call a parent, answer e-mail, and record the grades of one class.

- Set time limits to complete tasks and stick to them. Avoid starting, stopping, and restarting. For instance, do not start writing a quiz, stop halfway through to check e-mail, then return to the quiz. This wastes time.

- Keep your desk organized and all of your files in order.

- Have your students complete a do-now upon entering the classroom. You can take attendance while they work. The do-now helps settle the class, allowing for an efficient start to the math lesson.

- Establish practical classroom routines that promote efficiency. Make sure students understand and follow the routines; remind them of the proper routines as needed. (See "Establishing Efficient Classroom Routines" in Section Seven.)

- Monitor your students. Address problems quickly to keep the class on task and not waste time. (See Section Thirteen, "Managing Inappropriate Behavior.")

- Start meetings on time and keep focused on your agenda. A little small talk is fine during a conference with a parent or guardian, but it should not unreasonably extend the meeting or detract from the meeting's purpose.

- Encourage the help of your students whenever possible. Students can help pass out and collect papers, change bulletin boards, and straighten books on shelves.

- Strive to be organized. Being organized makes every day easier.

Time management is a major element of organization. Take control of time and you will "seize the moment" and work efficiently throughout the day.

The Importance of Effective Substitute Plans

It is the rare teacher who is *never* absent from class. Even if you manage to dodge cold and flu germs through the entire year, you probably will be out of class some days to attend workshops or seminars, or to address a personal obligation. As soon as you know that you will be absent, follow your school's policy to arrange for a substitute.

SUB PLANS THAT MOVE YOUR STUDENTS FORWARD

The ideal substitute for a math class is someone who is certified to teach mathematics. Unfortunately, most substitutes for math classes were not math majors and may not be knowledgeable in mathematics, particularly higher-level courses such as geometry, trigonometry, or calculus. Unless you know the qualifications of your substitute, you should avoid leaving sub plans that require him or her to teach unfamiliar material, a situation that will only frustrate everyone and result in little learning.

One option is to leave review material that gives students practice using and applying skills they have recently learned. Reinforcing material on current topics keeps students focused on what they have been learning and minimizes losing momentum in moving forward. Avoid leaving what is often referred to as "busy work," an assignment that has no relevance to the current topic or course, and one that students will immediately recognize as something merely to keep them occupied. Busy work has scant educational value and seldom keeps students busy.

When considering the types of plans to leave for a substitute, you should take into account the makeup and nature of your classes. Some students, for example, may work well in groups, even for a substitute, whereas others may need constant guidance during group work, which may be difficult for a substitute to manage. Leaving enough work that is appropriate for the abilities and attitudes of your students reduces potential problems.

You should impress upon your students that the work you leave for them is important. If students feel that the material assigned by a substitute will not be assessed, many are unlikely to apply themselves to its completion. Always leave directions for your substitutes to either collect work at the end of class or instruct students to finish the work for homework, which you will collect upon your return. When you return, be sure to collect and review the assignments.

THE TRUE EMERGENCY PLAN

Although you should try to leave substitute plans that are a part of your current unit of study, this is not always possible. We have all suffered that debilitating stomach virus that steals upon us during the night and makes it impossible to even think about designing a quick sub plan and phoning it in to a colleague. For these occasions, you should have an emergency sub plan prepared.

Your emergency sub plan should include material for at least one day, although you might consider having emergency plans for two days or more, depending on your school's policy. Because an emergency plan is prepared long before it might be needed, you should choose material that students can do at any time of the year. General reviews of math skills, self-correcting worksheets, sheets that require students to solve problems to decipher a message, or cumulative reviews provided in your textbook are possibilities.

Keep your emergency plans in a folder or large envelope. Clearly label them and store them in a safe place, perhaps a bottom desk drawer or cabinet where they can remain until needed. When necessary, you can instruct a colleague to give the plans to your substitute.

Because of their generic material, emergency sub plans should only be used in an emergency. Though they may not cover your students' current topic of study, these plans will provide a math activity that your students otherwise might not have.

LONG-TERM ABSENCES

If you know that you are to be absent for an extended length of time, notify your supervisor so that he has a chance to seek a substitute who is certified or at least knowledgeable in math. Perhaps you can meet with your substitute before you leave and discuss your objectives for your classes, and your routines and procedures. Maybe the substitute can observe your classes prior to taking over, thereby making a smooth transition for your students and substitute.

ESSENTIAL COMPONENTS OF A SUB PLAN FOR YOUR MATH CLASS

The requirements for substitute plans vary. Some schools have teachers follow a specific format. Others allow teachers broad latitude in how they write plans for substitutes. No matter what format you use, your sub plans should be thorough and clear. They should include the following essential components:

- ◎ Your name.
- ◎ Your schedule: Include room numbers, subjects taught, bell schedule, and a brief explanation of any duties you may have. If you team teach or share rooms with other teachers, include their names.
- ◎ Current seating charts for each class: Note students with special needs, for example, those with limited English skills or those who have medical concerns the substitute should know about. (Refer to your school's policies

about sharing student information.) Also mention any students who you feel can be relied on to assist the substitute.

- One reference sheet with vital information, including:
  - Location of classroom keys
  - Any special instructions, for example, a scheduled assembly
  - Procedures for distributing calculators if necessary to complete an activity
  - Procedures for using computers, if applicable
  - Location of classroom supplies such as pencils, extra paper, markers, or chalk
  - Procedures for restroom use and location of hall passes
  - Name of a colleague that the substitute could contact with questions
  - Instructions for dealing with students who misbehave
  - How to contact an administrator or the nurse in case of an emergency
- The assignment: Include thorough instructions. If worksheets are a part of the assignment, be sure to provide enough for all students. If students are to use books that are stored in your classroom, explain where the books are located. State whether the substitute is to collect the work or assign it for homework.

The best math classes move forward in learning every day. Leaving effective substitute plans helps ensure that your classes move forward even if you are not present.

Quick Review for Organizing for Success

To a large degree your success as a math teacher will depend on how well you organize your school day. The time you spend on organization will be some of the most effective use of your time. The following tips can help you focus on organization:

- Create and regularly update a personal calendar with important dates and events. Use your calendar for both short- and long-term planning.
- Use a binder to help you organize papers, and also to keep handy items such as pencils, markers, CDs, and flash drives.
- Establish practical routines for yourself and your students. Effective routines can save time.
- Prioritize tasks. Attend to the most important things first.

- Avoid neglecting minor tasks that, if permitted to accumulate, can become major ones.

- Organize your classroom for efficiency. Obtain sufficient storage space and store supplies and materials so that they are accessible.

- Encourage your students to accept a role in keeping the classroom organized. Students should follow classroom routines, return materials to their proper places, and keep work areas and aisles clean and uncluttered.

- Keep your paper and computer files organized. Set up files logically, placing all materials for a specific subject or topic in the same folder. Name computer files clearly.

- Manage your paper load. Daily attention to paperwork will prevent papers from accumulating.

- Use time efficiently to enhance your effectiveness as a teacher.

- Provide effective substitute plans to help ensure that your students are involved with math even though you are not present.

Teaching is a stressful profession. With all of your teaching and nonteaching responsibilities, there will be days you will be hard-pressed to complete all you set out to do. Effective organization and time management can help you make the most of every day.

SECTION SIX

Planning a Great First Day

The first day of class is filled with anticipation for everyone. The beginning of a new school year promises challenges and opportunities, demands hard work and persistence, and offers the rewards of personal growth and satisfaction.

While you may be wondering how you will meet the objectives of your curriculum, work within the drawbacks of a difficult schedule, and create meaningful math lessons for diverse learners, your students have their own concerns. Perhaps they are entering a new school where they will meet many new students from other parts of town; or maybe they had a poor experience in math last year and are convinced they cannot learn math; certainly, as adolescents, they are worried about what to wear, whether they will make new friends, and how they will manage to complete their math homework when they participate in after-school sports, take music lessons, or have dance class. Compounding all this is the fact that your students do not know you, and you do not know them.

First days have all of the elements for becoming stressful exercises that can undermine the rest of the year. Fortunately there is much you can do to make certain that your first day of classes is successful and sets a positive and enthusiastic tone for the year ahead. Careful planning and thorough preparation will help you manage the required first-day tasks and get your math classes off to a great start.

The First-Day Basics

No matter how long you have been teaching, it is normal to feel a little nervous and apprehensive on the first day of school. You will be meeting your classes for the first time, and along with the need to focus your students on math, there are numerous administrative details to handle. With so much to do, you must prioritize. Attending to the basics efficiently will enable you to make an easy transition to start your math classes.

Managing the basics for the first day actually begins the night before. Consider the following to start your first day smoothly:

- Plan what you will wear. Be sure your clothes are clean and set aside in one place. Washing a blouse or shirt during breakfast or having to search for that perfect pair of slacks is sure to make you feel rushed.

- Pack your bag or briefcase the night before with everything you need to take to school. Waiting to pack in the morning risks forgetting things or spending time looking for them.

- Go to bed a little early and get a good night's sleep.

- Wake up a little early to give yourself extra time; eat a solid breakfast.

- Prepare a good lunch (and be sure to take time for lunch).

- Plan to arrive at school early so that you have time to check in at the office and attend to any final preparations. Before the students arrive is a great time to complete any personal paperwork that is waiting for you in your mailbox. These might include emergency contact information, verification of your address and phone number for the school directory, payroll adjustments, parking decals, and so on.

The value of thorough planning cannot be overstated. Being prepared will help you to feel confident in meeting your students and starting the new school year.

IF YOU HAVE A HOMEROOM

If you have a homeroom, you have additional paperwork to do—attendance, insurance forms, parent-guardian emergency contact forms, lunch forms and menus, student schedules (if they were not previously mailed), locker numbers, student identification cards, student manuals, and health forms. In your school, homeroom teachers may be responsible for even more. To manage your homeroom duties smoothly, consider the following tips:

- In the beginning of the period, seat students in alphabetical order, leaving empty seats for absent students. Seating homeroom students alphabetically makes it easier to learn their names, and also makes taking attendance easier as empty seats indicate absent students. Furthermore, alphabetical seating facilitates handing out and collecting materials that contain students' names.

- Because it is unlikely you will know the students in your homeroom, it is beneficial, particularly during the first few days, to count the number of students in class as well as take attendance. This will ensure that everyone who is supposed to be in the class is in fact there. Be sure to follow your school's procedures for recording absences and tardies.

- When taking attendance, try to match names with faces. This will help you to learn your students' names more quickly. If a student is absent, call his name to make sure that he is not in the incorrect seat. If he is, remind him which seat has been assigned to him. (It is not unusual for students to forget their seats in the beginning of the year, especially if they have been assigned seats in eight or nine different classrooms.)

- Obtain a large folder or envelope in which to place your homeroom materials. This will reduce the chances of papers being lost or misplaced. Store any materials to be distributed as well as materials students have returned to you in the folder. When students are absent, write their names on their materials and give them to the students once they return.

- To keep track of materials returned by students, make a copy of your class list on a grid. Label items and materials at the top and simply put a check by students' names as they return the materials. Inform students of the dates materials are due. You may create a grid of your own, or use the "Record of Materials Returned by Students" that follows.

- When making announcements or when announcements are given over the public address system, remind students to listen. Even if the information does not pertain to some students, it may pertain to others. Moreover, if students are talking and not listening to announcements, they will invariably wind up asking you to repeat the information. If there is excessive talking, you may not even hear the announcements, which will result in unnecessary confusion.

- If you are responsible for taking a lunch count, consider having a lunch sign-up sheet. Place the sheet (with a pencil) on a desk or table near the front of the room, and instruct students to sign their names as they enter homeroom. Before reporting the count, make sure that everyone who needs to sign did sign. You can follow this procedure with other sign-up sheets as well.

- Distribute and collect materials as efficiently as possible. Consider having students help. There is usually someone in every class who is willing to hand out and collect materials. (Note: Although students can help you with many routine tasks, it is generally not advisable to have students take attendance. Remember, you are responsible for reporting which students are absent.)

- Be sure to adhere to your school's policy regarding the flag salute, and any other special procedures.

As in any other class, you need to establish practical rules and routines for homeroom. Consistency in following procedures will eliminate gray areas that might lead to confusion and help homeroom to set a positive tone for the day.

Record of Materials Returned by Students

Teacher's Name _____

	Names of Students	Materials That Have Been Returned					
1							
2							
3							
4							
5							
6							
7							
8							
9							
10							
11							
12							
13							
14							
15							

Record of Materials Returned by Students (continued)

Materials That Have Been Returned

	Names of Students						
16							
17							
18							
19							
20							
21							
22							
23							
24							
25							
26							
27							
28							
29							
30							

DOUBLE-CHECKING MATERIALS AND SUPPLIES

Before students enter the school, you should double-check all of the materials and supplies you will need for the first day. (See "Your Schedule and Class Lists" and "Setting Up Your Classroom" in Section Two, and "Basic Supplies, Materials, and Equipment" in Section Three.) Even if you have gathered and organized everything you need for your first day's classes (and homeroom if you are assigned one), you should make certain that everything is ready. This includes keys for the classrooms, class lists, seating charts, all handouts, books, workbooks, projectors, transparencies, markers, chalk, pencils, papers, erasers, materials for activities, and furniture in the classroom. A final check will reduce the possibility of surprises—for example, finding that a few student manuals have disappeared, forcing you to try to obtain some as students wait, or discovering that a few desks have been mistakenly removed from your classroom and placed in another. Attention to these kinds of details can eliminate the time-consuming tasks of searching for things you already had.

KEYS TO BEING CALM AND COMPOSED

Every teacher, even those with years of experience, feels at least a little anxious on the first day of school. To reduce any first-day nervousness you might have, try the following:

- Understand that being nervous arises from the anticipation of all you have to do on the first day. It is a normal reaction. Just about everybody else in the entire school is experiencing some of the same feelings. You are not alone.

- Have all of your materials and supplies ready, know your schedule, and plan great activities. Being prepared helps to boost your confidence and ease anxiety.

- Have more than enough math activities planned to fill each period. Knowing you will not run out of things to do reduces worry.

- If you are a veteran math teacher, draw on your experience regarding previous first days. Rely on effective strategies and improve on things you felt you could have done better.

- If you are a first-year math teacher, keep in mind that this is the first day of your career. Every teacher was a first-year teacher and made it through the first day. You will, too. Look forward to it with positive anticipation.

For most teachers, nervousness vanishes once they meet their students. As you become involved with your students, you will focus on the activities at hand and will progress capably through your day.

WELCOMING STUDENTS AT THE CLASSROOM DOOR

As students come to your class, welcome them at the door. If you teach in different classrooms throughout the day, attempt to get to the classroom ahead of your students so that you can greet them. Meeting students at the door with a confident smile shows that you are approachable and in charge; a friendly greeting can reduce any apprehension your students might have. In addition, by standing at the front door, you can see that students enter the classroom in a quiet and orderly fashion, which helps to ensure that the class begins on the right note.

To save time and help students find their seats, either project the class's seating chart on a screen, or tape the chart to the board at the front of the room. Orient the chart so that the top is aligned with the front of the classroom. As students enter the classroom, direct them to the chart. They can then find their correct seat. This eliminates the confusion of having students sit and then have them later move to their assigned seat. (See "Seating Charts" in Section Two.)

INTRODUCING YOURSELF

Once students are seated, introduce yourself. Be positive and upbeat during your introduction, which should be a genuine reflection of your expectation for the year.

While you should not share personal information on the first day, you should say your name clearly and write it on the board with your title (Mr., Ms., Mrs., Miss, Dr.). You might mention the math classes you are teaching this year, and why you enjoy teaching math (or this particular course). You might also tell your students how long you have taught in your school, or if this is your first year. If you have taught some of your current students' older brothers or sisters, say so and ask them to pass along your hello to their siblings. You might also explain where you went to college and why you like math. Be sure to mention to your students that you are looking forward to working with them.

Sharing a little about yourself encourages students to view you as a person to whom they can relate. It can help to put both you and them at ease.

GETTING STARTED

After introducing yourself, take attendance. As you call the names of your students from your class list, check your seating chart to make sure that students are in their correct seats. As it is likely that you will find it necessary to move some students to different seats, and new students may come into the class or some students may leave, explain that you may need to change seats in the future. Seating charts with assigned seats are valuable because they help you to learn the names of students quickly and prevent friends from sitting in the back of the room.

Before continuing, you should briefly discuss emergency procedures in accordance with school policy. For a fire drill, explain the procedure to be followed for exiting the building. If your school has a lockdown procedure in place, be sure you are familiar with it and inform students accordingly.

After these necessary requirements are satisfied, you are ready to move on to your math program. This should be the major focus of the first day.

Providing an Overview of Your Math Class

One of your most important goals for the first day should be to provide your students with an overview of your math class, including the following:

- A course description
- Your expectations for the class
- Classwork and homework
- Math projects or special activities
- Tests and quizzes
- Midterm and end-of-the-year examinations
- Grading policy
- Supplies that students will need

In your course description, you should share with your students the scope of your curriculum and the major topics you will cover. Including what your students will study and why they will study this material provides both direction and purpose for learning.

By sharing the expectations you have for the class with your students, you are informing them of what will be required of them as individuals. You may find it helpful to hand out copies of the following "Responsibilities of Math Students" to your students, or to adapt this list with your own requirements. The sheet identifies various behaviors that your students should demonstrate in your class.

Responsibilities of Math Students

Name _____ Date _____ Period _____

You will be more successful in math class if you accept the following responsibilities:

1. Report to class on time each day prepared to learn math.

2. Bring all of the materials you need for class, including your textbook, assignment pad, pencils, and math notebook.

3. Always complete your classwork, homework, and any other activities. Be sure to make up any work that you miss because of absences.

4. When you do not understand something, ask questions. It is likely that others have the same questions as you.

5. Work cooperatively with your classmates.

6. Be courteous and respectful to others. Do not disturb others.

7. Listen to and follow directions.

8. Be open to new ideas and learning.

9. Understand that succeeding at any worthwhile activity requires hard work and persistence. This includes math.

10. Always do the best you can. Do not allow yourself to believe that you cannot learn math. You can, but you must be willing to accept the challenge and apply yourself.

Your class overview should also discuss classwork, homework, projects, and special activities. Explain the types of classwork students can expect, as well as the type and amount of homework. For example, will students be expected to complete a do-now upon entering the classroom? Can they expect to receive homework every night, including Fridays and before holidays? Will students work in math groups? Will math projects be a part of your class, and if so, how many projects will be assigned? What topics in math and special activities might students look forward to? You might, for instance, plan to observe *pi* (3.14) day on March 14 (the fourteenth day of the third month) by presenting an activity about circles and conclude the class with eating a piece of pie. What about writing, which is an essential component of every math class? How much writing will students be required to do in your math class? Will they write in math journals, offer written explanations of the solutions to open-ended problems, or will their writing primarily appear in reports or projects? Explaining the kinds and amount of work students will be doing in your class can stimulate interest and reduce the anxiety of not knowing what is to come.

Along with daily work, your class overview should discuss tests and quizzes. About how often will tests be given? At the end of every chapter in the text? At the end of each unit? Will you surprise your students with pop quizzes? If so, how often? Once a week? Once every two weeks? Once in a while? Will your students be taking midterm and end-of-the-year exams in math? Will you use portfolios as assessment tools? When you discuss these kinds of questions, you ease students' apprehension and can help them adjust to your assessment requirements.

During your overview you should inform your students of the supplies they will need for your class. If they received a letter during the summer listing supplies they need, review the list. If your school does not provide student agenda books, consider suggesting that they obtain an assignment pad. Their supplies should also include a three-ringed binder. Using the binder for note taking as well as storing their tests, quizzes, and activity sheets reduces the potential problem of losing papers. Though a binder and pencils are basic supplies for every math class, your students may also need other materials, such as a ruler, protractor, compass, or markers to use at home. If your students need to buy a calculator, tell them which type of calculators your math department recommends. If students bring their calculators to school, be sure they label them with their names. White correction fluid or nail polish works well for writing names on calculators, because these substances do not rub off easily.

Although some teachers prefer to limit the overview of their math classes to expectations, description of the subject matter, required work, and supplies that students will need, others also include their grading policy. If you choose not to include your grading policy with your class overview, you should share it with students at a later date. It is only fair that students understand how you will determine their grades. (See "Devising a Fair System of Grading" and "Ways to Assess Student Learning" in Section Twelve.)

Providing your students with an overview of your math class introduces them to your course's subject matter and shares with them your vision of the class. When students know what to expect as well as what is expected of them, they are better able to move forward in learning math.

Learning About Your Students

If you are like most math teachers in middle or high school, you probably have between 100 and 150 students. Learning their names and getting to know them is, of course, essential to working with them effectively. However, this is not an easy task when you may have several classes with twenty-five or more students in each class. Distributing copies of the following "Facts About You" can help you compile student contact information and learn some of the things your students like and dislike.

Facts About You

Name _____ Date _____ Period _____

Home Phone Number: _____

Names of Parents or Guardians: _____

_____

Phone Number(s) of Parents or Guardians: _____

E-Mail Addresses of Parents or Guardians: _____

1. What do you like to be called? _____

2. Do you have any brothers or sisters? _____ If yes, list their names and state if they attend or have attended this school. _____
_____

3. What is your favorite subject? _____ Why is this your favorite? _____
_____

4. What is your favorite topic in math? _____ Why is this your favorite? _____
_____

5. What are your goals for the future? _____
_____

6. What is your favorite TV show? _____

7. Who is your favorite singer or musical group? _____

8. What hobbies do you enjoy? _____ Why do you enjoy these hobbies? _____
_____

9. What after-school activities or sports do you participate in?
_____
_____

10. What is your favorite sport and your favorite team? _____
_____

After collecting the worksheets, be sure to read them. Note an interesting fact or two about each student and mention it to them at some point during the year. Most students are surprised when you comment on something they included on the worksheet. They will be pleased that you are interested enough in them to remember something they wrote.

NAME CARDS

An easy way to learn the names of your students, and which also helps students learn each others' names, is the name card. Placed on desks, name cards help you to associate names with faces.

To make name cards, you will need white, unlined $8\frac{1}{2} \times 11$-inch sheets of paper and dark markers. Pass a sheet of paper out to each student. Tell students to hold the paper lengthwise. Starting with the bottom, they should neatly fold the edge up about two-thirds of the way to the top. They should then fold the top edge down to meet the crease created by the first fold. Their paper should now be folded into three equal parts, or panels. Pass around markers and instruct your students to print their names lengthwise on the middle panel. They should place their name card on their desk so that their name is at the front of the desk. Remind your students to bring their name cards to class until you—and everyone else in the class—have learned everyone's name.

CIRCLES OF ME: A GETTING-ACQUAINTED ACTIVITY

An interesting and enjoyable way to become acquainted with your students and help them get to know each other is to complete the activity "Circles of Me." Along with honing listening skills and requiring that students follow directions, this activity reviews some math terms and promotes thinking about math. It also reveals information about the students in your class: which students are vocal, which are quiet, who is prone to call out, and whether your class is one abounding with energy and enthusiasm or rather shy and reserved.

- ◉ Start the activity by explaining to your class that you would like to get to know them and at the same time review some math terms. Distribute copies of the worksheet "Circles of Me" on page 113. Instruct students to write the name they prefer you address them by in the circle located at the center of the sheet.

- ◉ Note the circle with the word *Positive*, and ask your students why they think this word is placed on the right side of the diagram. Students should realize that positive numbers are located to the right of zero on a number line, and that it can be assumed that zero is the center of the "Circles of Me" diagram.

Ask what the word *positive* means to nonmathematicians. Answers may include that it means certainty, confidence, acceptance, or good. Instruct your students to write something positive about themselves in the circle labeled *Positive*.

○ Follow this same procedure for the circle with the word *Negative*. Note that negative is to the left of the center, and that negative is the opposite of positive. Instruct your students to write something about themselves that can be improved in the circle labeled *Negative*.

○ Discuss the words *variable* and *constant*. Mathematically, a variable is a letter that is used to represent a number. The value of a variable may change, depending on the problem. In the circle labeled *Variable*, instruct your students to write something about themselves that may change or is changing. Unlike a variable, a *constant* is something that does not change. In the circle labeled *Constant*, instruct your students to write something about themselves that does not change.

○ Direct the attention of your students to the circle labeled *Symbol*. Explain that a symbol in mathematics is a sign that represents something—for example, an operation or a quantity. The + sign represents the operation of addition. In the circle labeled *Symbol*, instruct your students to draw a symbol that represents some aspect of their lives.

○ The last circle to complete is labeled *Rule*. Discuss that there are many rules in mathematics. For example, the order of operations states that expressions within grouping symbols must be simplified first. In the circle labeled *Rule*, instruct your students to write a rule that they follow in their lives.

Note that depending on the level and abilities of your students, you may prefer to change some of the terms in the circles. You may also add circles or delete some.

After students have completed the worksheet, allow them to conference in small groups. Have each student choose at least one circle of another student and use the information in the circle to introduce that student to the class. An option is to have each student select at least one circle of his or her own and introduce him- or herself.

You should also complete this activity and share your results with the class. As you get to know your students, they will get to know you.

At the end of the activity, collect all of the worksheets and be sure to read them later. The "Circles of Me" will help you learn much about your students.

Circles of Me

Name _____ Date _____ Period _____

Symbol

Variable

Positive

Your Name

Negative

Constant

Rule

Providing a Math Activity on the First Day

Even though you usually have plenty to do on the first day of classes, you should provide time for a math activity. The benefits are significant. A math activity on the first day emphasizes the importance of math in your class, gets students working on math, and helps to reinforce the idea that students are expected to work in class.

There are many activities and problems you can choose for the first day. We recommend that you select those that are nonthreatening, but which will stimulate student thinking. Perhaps you would like to use the following first-day number puzzler.

Number Puzzlers

A number puzzler is a fun way to get your students thinking about and working on math right from the start of school. You will create the problem, for which the answer will be the numerical equivalent of the month and day on which your first class meets. Suppose that your first day of school is September 5. The answer to the puzzle, therefore, is 905, which represents the ninth month and fifth day.

Three examples are provided for September 5. Use the examples to create a puzzler of your own for your first day of school. Be sure to base the puzzler on the general grade-level abilities of your students. When students have found the answer to the puzzler, ask them to look at the number and consider what the number represents. You may offer a hint that the answer has something to do with the calendar.

Example 1. The hundreds digit is equal to 3 squared. The tens digit is the only number that is neither positive nor negative. The units digit is the second odd prime number. What is the number?

Example 2. Divide 500 by 10 and add the quotient to the product of 30 times 30. Then subtract the product of 9 and 5. What is the number?

Example 3. The hundreds digit is the solution to the equation $0.5x = 4.5$. The tens digit is the same as the slope of a horizontal line. The units digit is equal to 40% of $(3 \times 4 + 0.5)$. What is the number?

If time remains, you may encourage students to make up number puzzlers of their own. They can exchange their puzzlers with a partner.

By presenting a number puzzler for the first day, you will stimulate the thinking of your students in an enjoyable activity. If you intend to use the same puzzler later in the day, ask students not to reveal the answer to friends in other classes.

Handing Out Texts, Workbooks, and Other Materials

If you have time on your first day of classes, you may distribute texts, workbooks, and other materials to your students. This is particularly true if you are teaching in a block schedule or have an extended period for math. If, however, your class is the typical forty-five minutes, you may prefer to hand out texts on the second day of school.

Whenever you hand out texts and related materials, always have an activity for your students to complete. Perhaps you can distribute the reproducible "Facts About You," or ask students to create number puzzlers that they can share with each other. (Both are described earlier in this section.) Yet another option is to distribute and instruct students to complete a skills inventory while you pass out the books. Most texts have skills inventories that are designed to determine the strengths and weaknesses of students at the beginning of the year. Of course, you may prefer to design your own inventory, based on your curriculum. Skills inventories are useful for identifying the general abilities of your students.

When handing out any materials to students that are to be returned, you must keep accurate records. For example, when distributing textbooks, number each text and write the number next to each student's name in your record book. Double-check that the name and number are correct before moving on to the next student. If your school requires students to write their names in their texts, tell them to print their names in pen. If a lost book is later found and the student's name is written in it, the book can be easily returned. At the end of the school year, students should be required to return the same book they were given at the beginning of the year.

After handing out texts, review the books with your students. Briefly go over the contents together, noting the major topics you will be studying. Also point out special sections of the book—such as lists of study tips, articles that connect math to other subjects, extensions, chapter summaries, test reviews, tests, projects, glossary, index, and any special features. Most texts provide a wide assortment of activities and reference materials. Emphasize that a math text is far more than just a book of math problems.

Quick Review for Having a Great First Day

The first day of school can be frenetic and overwhelming, but it can also be an efficient and enjoyable start to the school year. Keeping in mind the following can help ensure that your first day is a wonderful experience for you and your students:

◎ Start planning well in advance of the first day.

◎ Make sure that all your supplies and materials are ready.

- Plan math activities that will stimulate your students and help focus them on the year ahead.

- Get a good night's sleep before the first day, eat a solid breakfast, and arrive at school early to handle any last-minute details.

- Attend to paperwork as efficiently as possible.

- Remember that everyone is nervous the first day, but also realize that effective planning and preparation can reduce apprehension and build confidence.

- Welcome your students, and set a tone conducive to learning.

- Offer an overview of your math class, clearly informing students of what you expect and what they can expect.

- Start learning the names of your students and getting to know them.

Make the first day a positive event for yourself and for your students. A positive first day is the best beginning for a productive and rewarding school year.

SECTION SEVEN

Managing Your Math Classroom

The underlying foundation of a successful math class is effective classroom management. A well-managed class operates smoothly, transitioning from the opening of class through the lesson and activities to the end of class. Throughout the period, students are involved in exploring math concepts and mastering math skills. The class is a center of activity and learning.

Managing your math classes effectively requires planning, consistency, and diligence. You need practical and efficient routines that support your goals and foster student progress. You must establish and enforce clear rules that ensure an orderly classroom, promote productive work habits, and enable students to excel. The best math classes, without question, begin with effective classroom management.

Establishing Efficient Classroom Routines

Routines provide a framework for conducting the necessary daily activities that support the efficient operation of a class. Once routines are established, students know what is expected of them and they can assume more responsibility for their learning.

Although particular routines will vary among math teachers, several—including taking attendance, dealing with students who arrive to class late, handling requests to leave class, distributing and collecting materials, and utilizing calculators and computers—are basic to just about every math class. The way you handle such routines is a major factor in the development of a positive and efficient classroom environment.

TAKING ATTENDANCE

Set up your attendance book (hard copy or electronic) before school begins and also at the start of each marking period. Note days when school is not in session or when classes will be missed because of assemblies or special events. Consistently update your records to account for days that school is closed because of inclement weather or emergencies.

Take attendance at the beginning of class each day. Having students work on a do-now as soon as they enter the classroom while you take attendance helps them get settled and focus their attention on math.

Attendance records must be accurate. Be sure to follow your school's attendance policy.

DEALING WITH STUDENTS WHO ARRIVE LATE TO CLASS

As with attendance, follow your school's policy regarding late arrivals to class. If there are disciplinary measures to enforce, be sure to enact them. Avoid slipping into the habit of permitting students to come to your class late. This will only encourage tardiness and undermine the importance of your math class.

When a student arrives late, acclimate him to what is going on with a minimum of disruption. Perhaps a student next to him can quietly explain the current activity. Discourage questions such as "What did I miss?" or "Did you correct homework?" at this time. Explain to the student that you will be glad to answer any questions he may have after class; addressing such questions when the student arrives will take you off track and disrupt your lesson. (See "Habitual Lateness to Class" in Section Thirteen.)

SMOOTHLY HANDLING REQUESTS TO LEAVE THE CLASSROOM

Every teacher must deal with student requests to leave the classroom. Most requests are legitimate—for instance, the ill student who needs to see the nurse, the student who has an appointment with a guidance counselor, or the student who must use the lavatory. The purpose of some requests, however, is simply to get out of class, perhaps to meet a friend in the hall or to enjoy a change of pace.

You should follow your school's policies for students leaving classrooms. If there is no definitive policy, establish one for your classes. Consider a policy in which students cannot ask to leave class during instruction, except in the case of an emergency. When students are permitted to leave class during instruction, they may miss valuable explanations and upon their return, they may ask questions about concepts you have already explained. This slows the pace of the class, and may bore students who understand the material.

To help manage students leaving class, consider using a sign-out sheet such as the reproducible "Classroom Sign-Out Sheet" in Section Two. A sign-out sheet provides you with a record of times and places students go. In the event of an emergency, or if the office calls for a student who has left class, you will know where she is. Retaining and periodically reviewing sign-out sheets allows you to note patterns of students leaving the classroom. Attach the sign-out sheet to the board with tape, or place it on a table near the door. Having a pencil with the sheet eliminates the chance of students disturbing classmates to borrow one. Students should write their name, the time they leave class, their destination, and the time they return to class on the sheet. Consider permitting only one student to leave class at a time, again, except in the case of an emergency.

Many schools require students to carry a hall pass when they leave class. Hall passes help to ensure that students go where they are supposed to go. Keeping returned passes on file enables you to confirm that a student arrived at her destination. If your school provides passes, be sure that students do not leave class without the proper pass. If your school does not provide passes, consider using the reproducible "Hall Passes" in Section Two. (See "Repeatedly Requesting to Leave Class" in Section Thirteen.)

DISTRIBUTING AND COLLECTING MATERIALS

Because distributing and collecting materials can consume a significant portion of class time, you should establish routines that streamline the procedure. In addition to handing out and collecting the typical paperwork of class—homework, classwork, tests, and quizzes—you may need to hand out and collect a host of items, including calculators, rulers, protractors, and manipulatives. Following are some suggestions:

- If possible, put materials students will need for the day's work on their desks before they come into the class. This saves time and permits you to start your lesson sooner. (Unfortunately, few math teachers enjoy a schedule that allows them to do this.)

- Assign trustworthy students to help you hand out and collect materials. To speed the procedure, you might place the materials to be distributed at the front of the room before students enter. As soon as students come into the classroom, your helpers hand out the materials. This system will work for papers and items such as rulers and protractors. Students can also help distribute manipulatives. When collecting materials, instruct student helpers to place items in specific folders, trays, or boxes.

- Instead of having students hand out materials, you might place papers or materials on a table at the front of the room and instruct students to pick them up as they enter class.

○ Emphasize to your students that they should always return materials and equipment to their proper places.

○ Calculators require special consideration in your math class. The following can help you develop routines for the use of calculators:

- If your school requires students to obtain their own calculators, inform students at the beginning of school (or via a letter to their parents or guardians prior to the start of the term) of the model and type of calculator they should purchase. This helps to ensure that everyone has the same calculator, making instruction easier. Having students write their name on the back of their calculator in nail polish or white correction fluid increases the likelihood that lost or misplaced calculators will be returned to their owners.

- If you have calculators that must be distributed and collected during each class, you must maintain accurate records to ensure that they are not lost. Number each calculator, using nail polish or correction fluid. If different math classrooms have their own sets, write room numbers on them in addition to the calculator number. Record this information, as well as the serial number of each calculator.

- Assign a specific calculator to each student in your classes. Explain to your students that each day they should make sure they use the correct calculator and that they return it at the end of class. Printing a class roster with the number of each student's calculator next to his or her name and posting it on the board allows students to double-check their calculator's number. This eliminates the need for students to ask you for the number of their calculator in the event they forgot it. Also encourage students to write their calculator number in their agenda or math notebook. Emphasize that students are responsible for the calculator that was assigned to them.

- Keep the calculators in a storage box or bin, with about ten calculators in each, and place the boxes near the front of the room. Store calculators 1–10 in Box 1; 11–20 in Box 2; and 21–30 in Box 3. Writing the numbers of calculators stored in each box makes it easy for students to find specific calculators and prevents students from having to sort through all three boxes. Instruct students to pick up their calculators upon entering the classroom.

- At the end of class, consider having three students (one for each box) collect the calculators and place them in their proper boxes.

- You should always count the calculators before class and recount them at the end of class before students leave. Do not dismiss the class until all calculators are accounted for. If any calculator is missing, you can quickly find it and the student responsible for it.

- Place the boxes of calculators in a locked closet or cabinet before leaving class.

PROCEDURES FOR STUDENT COMPUTER USE

Whether you have computers in your classroom, or you must use a portable computer lab, student use of computers requires clear and consistent routines. If your school has a policy for computer use, be sure that you follow to it. If, however, your school lacks a clear policy, consider these suggestions:

- You should maintain a record of student use of computers. Consider using the following reproducible "Computer Sign-Up Sheet" or use it as a model to design a sheet of your own. Place a copy of the sheet at each computer in your class, and instruct your students to write their name, login time, and logout time. A sign-up sheet enables you to monitor computer use and reduces chances for vandalism.

- Consult the computer technician in your school regarding the routines for logging on and off the school's computer network. Follow her instructions regarding user names and passwords. Maintain a copy of each student's user name and password in case students forget. Keep this information confidential. No student should have access to the user name and password of another student.

- Ask the computer tech to install virus protection and Internet security software on your computers, and to create measures to prevent students from being able to change settings. Also ask her to block students from accessing personal e-mail and inappropriate Web sites.

- Instruct your students to practice good computer etiquette. Distribute and discuss the reproducible "Computer Etiquette," found on page 124.

Computer Sign-Up Sheet

Computer #_____   Room #_____

Date	Your Name	Login Time	Logout Time

If a computer is not working properly during class, attach a note to its monitor, "Do Not Use," to inform other teachers or students who use the room after you that this computer is down. You should contact your tech person immediately, explaining the problem and how you tried to resolve it. You may also note the problem on the reproducible "Record of Used Supplies or Malfunctioning Equipment" in Section Five.

Practical classroom routines provide a structure for you to teach and your students to learn math. Having procedures in place eliminates questions about what, when, and how to do things, freeing both you and your students to focus on the math lesson.

Computer Etiquette

Name _____ Date _____ Period _____

Follow the guidelines below when using computers in math class.

1. Computers should be used for math only.

2. Note your login and logout times on the sign-up sheet.

3. Follow your teacher's instructions.

4. Remain on task.

5. Visit assigned Web sites or sites that will provide you with the information you need to complete your assignment. Do not visit other Web sites.

6. Ask your teacher before downloading any information.

7. Tell your teacher about any hardware or software problems you encountered.

8. Avoid improper shutdowns.

9. Clear paper and other materials from the area around the computer before you leave.

10. Return the screen to the desktop when you are finished using the computer.

Achieving a Smooth Flow of Classroom Traffic

The movement of students in your classroom should be minimal, orderly, and purposeful. Permitting unnecessary or uneconomical student traffic may lead to wasted class time, interruption of your lesson, or distractions for other students. You can achieve a smooth flow of classroom traffic by establishing routines that guide students in entering, moving about, and leaving your room.

Think about the usual activities your students do in your classroom and set up the room to facilitate the movement necessary to these activities. Furniture should be arranged so that students can move around the room easily, yet see the board, screen, and you clearly. (See "Arranging Furniture to Enhance Math Learning" in Section Two.) Aisles should be clear of materials and wide enough for students to walk through. Materials and supplies, such as paper, staplers, and pencil sharpeners, should be placed where students can access them without disturbing others. For example, placing paper on a windowsill ledge where it is accessible but does not result in traffic blocking a view of the front board is a good idea. Positioning a waste basket at the back of the room where you are less able to monitor students discarding paper is not.

In addition to setting up furniture and positioning materials and supplies in a practical manner, consider the following ways to encourage a smooth flow of classroom traffic:

- ◎ Require students to enter class calmly and quietly. Standing at the door and greeting them as they step into the classroom serves as a subtle reminder that they are coming into your math class and need to behave appropriately.

- ◎ Requiring students to begin a do-now upon entering the room encourages them to go directly to their seats, reduces congestion, and can effect a quick, orderly start to class.

- ◎ If students are to use calculators for class, place the calculators near the door and instruct students to pick them up before going to their seats.

- ◎ Discourage students from getting out of their seats to sharpen a pencil, discard paper, or ask to leave the room while you are teaching or giving directions.

- ◎ Decide upon rules for students leaving their seats during class. For example, are you comfortable with students accepting the responsibility of getting up from their seat to sharpen a pencil? Or do you prefer that they always ask your permission before leaving their seat? Sometimes it can be more disruptive for students to ask rather than simply and quietly sharpening their pencil at an appropriate time (not while you are teaching).

- ◎ Before students leave the room, they should ask your permission. They should sign out (if you maintain a sign-out sheet) and obtain a hall pass if required. They should leave the room quietly and return quickly. Permit only one student out of the room at a time, except for an emergency.

○ Any students who help you hand out or collect material should be responsible and efficient. Avoid permitting helpers to socialize unnecessarily. Their attention should be focused on their task. Changing volunteers periodically is fair and allows different students to assume a role in class procedures.

○ Limit the number of students who write problems on the board so that they have enough room to work and that crowding at the front of the room is reduced. As soon as students complete their problems, they should return to their seats.

○ If students rearrange their desks for group work, you must establish an orderly procedure. Students should move their desks into squares or rectangles with minimal noise and confusion. At the end of the class, desks should be returned to their original arrangement.

○ Traffic around your desk should be limited. Your desk is your workspace. Students should not sit at your desk, use your computer, or have access to any materials in or on your desk.

○ Just before the end of class, you should inspect the room. Remind students that all materials should be put back in their proper places, the floor should be clear of papers and books, and desks should be straightened.

○ Students should not leave the class until you dismiss them and, once dismissed, they should exit the classroom quietly.

A math class without clear routines for student movement will lack focus in comparison to one in which practical routines are implemented. You should establish the routines that support your program and foster an efficient classroom.

Creating a Productive Math Class

One of your most important classroom management goals should be the promotion of productive work habits in your students. As students acquire sound work habits, they become more involved in the class's activities and procedures, require less guidance from you, and master math concepts and skills more quickly. They will become more independent learners. You should promote productive work habits from the start of each class to its end.

BEGINNING CLASS WITH A MATH DO-NOW

A do-now is a problem that students begin working on as soon as they enter the classroom. Math do-nows, which you can write on the board or project on a screen before students arrive, usually take about three to five minutes to complete, perhaps a little longer. They may take a variety of forms, including:

- Review of previously learned skills
- Summary of prerequisite skills for the day's lesson
- A math puzzle
- A tip for test preparation
- A math journal entry
- A word problem that may be solved in a variety of ways

Do-nows provide several benefits. They encourage students to come to class ready to work. The problems may reinforce past learning, supplement current instruction, or introduce new material. Do-nows also support your classroom management. While students are completing a do-now, you may take attendance, hand back papers, or circulate around the room making certain everyone is focused on her work. Rather than losing the first few minutes of class settling students and directing their attention to math, a do-now enables you to use the full period on instruction and activities.

There are many sources for do-nows, including many resource books. (You may want to consider *Math Starters! 5- to 10-Minute Activities That Make Kids Think, Grades 6–12*, by Judith and Gary Muschla, Jossey-Bass, 1999, and *The Math Teacher's Problem-a-Day, Grades 4–8*, by Judith and Gary Muschla, Jossey-Bass, 2008.) Your math text is also a likely source, particularly the notes in the teacher's edition or sections on supplemental or extended activities or problems. Newspapers can be a source of do-nows on current topics and themes, and the Internet can be a source of problems on just about anything. A search using the terms "math do-nows" or "math starter problems" will result in helpful Web sites.

By engaging students with a problem as they enter the classroom, math do-nows offer an effective way to start your classes. They send a subtle message that students are to spend all of their class time on math, set a positive tone for the lesson that follows, and foster productive work habits.

AGENDAS OR ASSIGNMENT PADS

Because they provide a place for students to write down the work they must do, agendas or assignment pads are essential tools for your students. Agendas and assignment pads are useful not only for students, but also for parents and guardians who help their children with schoolwork.

Instruct your students to always bring their agendas or assignment pads to class. They should write down their assignments, including page and problem numbers, any special directions, and when the work is due. All assignments, including homework, studying for tests or announced quizzes, and long-term projects, should be recorded on their agendas or assignment pads. Encourage your students to refer to their agendas or assignment pads often, especially before leaving school at the end of the day.

Depending on your students, you may want to reserve a few minutes near the end of class for them to copy down the assignment from the board. For some classes, you may find it beneficial to circulate around the room to make sure that students are being thorough in recording their assignments.

CLASSWORK

Each day your students will be engaged in various activities in your class. They may be listening to your explanation of a new concept, taking notes, exploring manipulatives, working in groups to solve multistep problems, or using technology to investigate mathematical applications. You, of course, will be interacting with them throughout the period. Providing your students with guidelines for classwork can help keep them on task and develop positive work habits.

Here are some suggestions:

◎ Always write the math lesson's objective on the board or screen. When students understand the purpose of the lesson, they will know what they are expected to do and are more likely to become involved with the material.

◎ Require students to use a consistent heading on all their papers—for example, name, date, period, and assignment on the upper right-hand side of the sheet.

◎ Require students to do all work in pencil.

◎ Require students to show their work in solving problems. Answers that do not show work are not acceptable.

◎ Require students to write legibly and keep papers neat and clear of smudges and big eraser smears. Explain that they should be proud of the work they submit and that their work reflects on them.

◎ Provide clear directions and explanations for all activities.

○ Plan a variety of activities that include demonstrations, manipulatives, technology, problem solving, applications, and modeling. Varying activities stimulates student interest and prevents boredom.

○ Mix individual and group work to provide different learning situations.

○ Engage all students in the class's activities. Teach to all parts of the room. Call on different students to answer questions, and avoid relying on the same students to assist you in class. Everyone should be given a chance to participate; everyone should be encouraged to contribute.

○ Circulate around the room as students work. If necessary, remind them of appropriate behavior and to remain on task. Answer questions they may have and offer encouragement and guidance, but avoid giving solutions to specific problems. Students benefit most when they solve problems themselves.

The most engaging math classes include a variety of activities for individuals and groups. Classwork expands the horizons of your math class and should be an integral component of your program. (See "Assessment Through Classwork" in Section Twelve.)

GROUP WORK

When students work in groups to solve problems, they learn both math and social skills. Group work fosters investigation, inquiry, and discussion, providing students with greater experiences and opportunities for learning than when they work individually. Group work builds the confidence of students, encourages critical thinking, and leads to new insights about mathematics.

Though you may base your groups on readiness, interests, or learning styles, often the best math groups are those that are formed randomly with a mix of genders and ethnicities. Base the number of students in your groups on the complexity of the problem to be solved. Whereas pairs or groups of three will work for problems that require some investigation and a few steps to solve, groups of three to five students are a better choice for complex problems that require significant research, analysis, and organization of data and testing of possible solutions.

A simple way to create random groups is to count down your roster in sets of the number of students of each group. For groups of four, for instance, give each student a number ranging from one through four. All number ones form a group, all number twos form another group, and so on. Before announcing the groups to your students, review them to make sure that the random set-up did not result in potentially troublesome combinations. For example, in most cases, best friends should not be in the same group, nor should two students who do not get along. Adjust the members of your groups as you feel necessary. Random formation loses its value if it results in unworkable groups. You should also change groups periodically to give students a chance to work with others whose viewpoints may present new and different problem-solving approaches.

For any group to work effectively, students must be willing to cooperate. They may assume various roles, including the following:

- The **group leader** guides the group to the goal of solving the problem. She makes sure that everyone contributes to the group and keeps the group moving forward.

- The **recorder** writes down notes of the group's ideas, strategies for solving the problem, and possible solutions. The recorder frequently summarizes what the group has done as it moves forward to solve the problem.

- The **checker** reviews the work of the group. Checkers make sure that the math used in solving the problem is correct.

- The **materials monitor** keeps track of the materials and supplies the group uses. He is responsible for obtaining and returning items to their proper places.

- The **time monitor** keeps track of time and helps the leader keep the group moving forward.

All members of the group, of course, are responsible for suggesting strategies for solutions, making conjectures, helping to gather and sort data, and testing solutions. All should accept responsibility for the group's success. To help your students work cooperatively in groups, distribute copies of "How to Work in a Math Group," which follows.

It is during group activities that some of the most important learning occurs. The work that goes on in your classroom is a major part of your overall math program. (See "Assessment Through Group Activities" in Section Twelve.)

How to Work in a Math Group

Name _____ Date _____ Period _____

The following suggestions can help you and the other members of your math group work together.

1. Accept responsibility for your behavior.

2. Be willing to work with all other members of your group.

3. Think about your ideas before speaking.

4. Share your ideas with your group.

5. Speak clearly, but quietly, when you explain your ideas. Remember that other groups are working, too.

6. Listen politely and consider the ideas of others before responding.

7. Ask for clarification if you do not understand someone's ideas.

8. Discuss disagreements calmly. Sometimes discussions over disagreements lead to even better ideas.

9. Stay on task.

HOMEWORK

Homework is an essential component of any math class. Meaningful homework can reinforce skills learned in class, connect learning to new skills, and develop positive work habits. As they complete their homework, students often master and apply skills that they studied in class but did not fully understand.

The homework you assign should always be purposeful and based on clear objectives. Unless it is designed to be a challenge, where students are to acquire new skills through discovery, homework should be an extension or review of the day's activities in class. Avoid assigning homework that requires skills for which your students have not been prepared. Such assignments will only result in frustration for you and your students.

Although your homework policy should be aligned with the homework policy of your school, especially the policy of the math department, you will probably have some latitude in the type, frequency, and amount of homework that you assign to your students. Some teachers, for example, assign homework that reinforces the skills taught in class, whereas others use homework to challenge their students by applying what they learned to new situations. Some teachers prefer to assign homework every night except Fridays and before holidays; others assign homework every night. Some try to assign roughly the same amount of homework each day, perhaps twenty to thirty minutes; others assign as much, or as little, as they feel students need to master the concepts and skills taught in class. You must decide how you will incorporate homework in your classes, including procedures for going over homework with your students to make sure they understand the material. You must also determine whether you will collect homework each day, grade homework, or simply check that students have completed it. (See "Assessment Through Homework" in Section Twelve.)

To help you assign homework that supports learning in your math classes, you should ask yourself questions such as the following:

- What is the purpose of this assignment?
- Is it relevant to the students?
- Does the assignment support the objective and math lesson presented in class today?
- How many problems should I assign?
- Do my students possess the skills necessary to complete this assignment?
- What types of materials or information will my students need to complete this assignment?
- When will this assignment be due?

Whenever you assign homework, you should write the assignment on the board or project it on a screen. Include page numbers and problem numbers, and any special directions. If necessary, explain the assignment so that students understand precisely what they are to do. Remind them to refer to their notes and classwork should they have trouble with their homework. If your school has a homework Web site or homework hotline, regularly update assignments and any special instructions. If you have space, you may also include the work that was done in class as a prerequisite for the homework. Posting assignments in this way is also helpful to parents and guardians who may be working with their children. These steps will also reduce the problem of students coming to class and complaining that they were unable to complete their homework because they did not know what to do.

Your school's homework Web site and hotline are also great places for posting information about upcoming tests, quizzes, long-term assignments, and math projects. Briefly describe each. For tests and quizzes, include any reviews you have distributed and materials that students should use to prepare. For long-term assignments and projects, include materials and handouts students have received. Always include due dates.

Few students enjoy homework; however, homework is the extra practice many students need to master math concepts and skills, helping them to connect math to other areas and apply it to real-life situations. Diligently completing homework helps build confidence and positive work habits that serve students well beyond your math class.

MATH JOURNALS

Math journals provide an opportunity for students to develop mathematical thinking and competency in communicating ideas about math. Journal writing requires students to reflect upon math and review what they have learned, leading to a greater understanding of math concepts and skills. When you read your students' journals, you will likely gain valuable insights into their overall progress in math.

The following suggestions can help you utilize math journals in your classes:

- A math journal should be a spiral notebook or composition book used exclusively for the journal. If the notebook becomes full, the student should start a new one.
- Students' names, course title, and period should be written on the cover of their journals.
- Each journal entry should be dated.
- Students may write in their journals at school and also at home.

- Encourage your students to write about math topics and problems that interest them; however, you may assign topics for students to write about, such as the following:
  - Questions they had and solved regarding specific problems
  - Specific problem-solving strategies
  - A particularly challenging problem and how they solved it
  - Insights about math
  - Explanations of solutions to specific problems
  - Discovery of a new concept
  - Musings about math—things they wonder about

- Students should develop their ideas logically and express them clearly. They should be specific and include details in their writing that demonstrate sound mathematical reasoning.

- Entries should be thorough but concise. Most entries should require about ten to twenty minutes to complete.

- Journal writing should not be done every day. Once or twice per week is usually sufficient for students to reflect about math without overburdening them.

- Encourage your students to review their journal entries periodically. They may find topics that they can revisit and write about from new perspectives or because they learned new facts.

- Be patient. Students will soon acquire the art of journal writing.

To help your students find topics for writing in their math journals, consider handing out copies of the following list of "Math Journal Writing Prompts." Feel free to use the questions on the reproducible as examples to create questions of your own that support specific objectives.

Math journals enable students to think about math from the perspective of their personal experience and not merely as rules or steps that must be followed. Journal entries become a record of a student's progress in the study and learning of mathematics.

Math Journal Writing Prompts

Name _____ Date _____ Period _____

Use the following questions to develop ideas for writing in your math journal.

1. What strategies did you use for solving this problem?

2. What new idea did you learn in math during this class?

3. Why was this problem easy? Or why was it hard?

4. What steps did you use to solve this problem?

5. What advice could you give someone for solving this problem?

6. In what other ways might this problem be solved?

7. What math shortcuts might you share with others?

8. What tip for studying for a math test can you give to others?

9. When have you used estimation?

10. What is your favorite topic in math? Why?

11. What topic in math would you like to know more about? Why?

12. When did you use math in another subject? How did you use it?

13. How do you use math outside of school?

14. How do you use computers in learning math?

15. Why do you like (or dislike) math?

MATH NOTEBOOKS

Every math student should have a math notebook. Though these notebooks may take many forms, a good choice is a separate three-ring binder with sleeves on its inside front and back covers, which provide additional storage space for your students' math papers.

All of a student's math work should be kept in the notebook. This includes do-nows, classwork, notes students take, homework assignments, activities, quizzes, and tests. All written work should be done legibly in pencil and dated.

There are many ways to organize math notebooks. One possible method is for students to keep classwork, homework, and activity sheets in the sleeve on the inside front cover, placed according to date. Tests and quizzes may then be stored in the sleeve on the inside back cover, again according to date. All other work should be kept in the binder and dated. When they keep papers in their notebooks, students may easily refer to past work. The notebook becomes a record of their progress.

To help your students organize and maintain math notebooks, you might distribute copies of the following "Tips for Keeping a Math Notebook," or use the tips as the basis for creating your own guidelines. Instruct your students to keep this sheet in their math notebooks for reference. Maintaining an organized math notebook is a valuable study aid and an important step to developing productive work habits. (See "Assessment Through Math Notebooks" in Section Twelve.)

Tips for Keeping a Math Notebook

Name _____ Date _____ Period _____

A math notebook contains a record of all the math you have done this year. Use it to store and review your work.

1. Obtain a three-ring binder with sleeves on the inside front and back covers to use as your notebook. Only math should be kept in your math notebook.

2. Date all work.

3. Always use pencil and write neatly in your notebook.

4. Start a new page of notes for each day of class. Notes should begin with the day's objective. They should include any diagrams, examples, formulas, and instructions. Use a highlighter to underline the most important information, such as formulas and rules.

5. Each homework assignment should begin on a separate page. It should include the page numbers and problem numbers, and have the proper heading with name, period, and date.

6. Use your notebook to store the math work you complete in class. This includes do-nows, notes, classwork, and activities. Organizing your work by date makes it easy to find specific assignments and activities.

7. Place classwork, homework, and activity sheets in the sleeve of the inside front cover of your binder. Keep papers unfolded and face up.

8. Place tests and quizzes in the sleeve of the inside back cover of your binder. Organize them by date and keep them unfolded and face up.

9. Keep your notebook clean and free of smudges.

10. Refer to your math notebook when you do your homework.

11. Use your math notebook when you study for tests or quizzes.

12. Bring your math notebook to class each day.

PROCEDURES FOR MAKING UP MISSED WORK

Because few students achieve perfect attendance during a school year, procedures for making up missed work in your class are necessary to ensure that students do not skip over important concepts and skills. Some students will assume the responsibility for completing assignments they missed, demonstrating an important aspect of good work habits, but others will need reminders from you. Clear and practical procedures provide students with direction for making up missed work.

If your school or math department has policies regarding makeup work, base your classroom policies on the established guidelines. If there is no formal policy, or the policy in place is general, develop one of your own.

Consider the following suggestions:

◎ All students must be required to make up all work they miss. Excusing some will undermine your rules.

◎ Establish clear guidelines and due dates. For example, if a student is absent one day, he must complete the makeup work one day after he returns. If he is absent two days, he has two days to complete the work. If he is absent three days, he has three, and so on. Students who miss significant work because of extended absences should meet with you and together you should set a reasonable due date for missed work.

◎ If students know in advance that they will be absent, they should inform you so that you may discuss any special instructions for assignments with them.

◎ Write each day's assignments for each class on a large Makeup Work Calendar that you can display on a bulletin board. When students return to class after being absent, they can check the assignments for the days they were absent. Be sure to keep your calendar current. Consider enlisting the help of a student volunteer for this.

◎ Keep extra worksheets for makeup work in a tray on a table or the windowsill. If you use a calendar for makeup assignments, place the tray near the calendar.

◎ Have set times, perhaps in the morning before the school day or after school, that students may see you regarding makeup work.

◎ Suggest methods for students to obtain assignments they miss while they are absent, including:

- Use the buddy system. Students may check with a reliable classmate who can provide them with the assignment they missed.
- Ask a friend or sibling who attends your school to see you after school, if possible, to obtain the assignment and materials. Suggest to your students that if they plan to do this, their friend or sibling should come to see you before school starts so that you may have the assignment ready for them after school.

- Check the assignments for math on your school's Web site or homework hotline.

○ Suggest methods for students to obtain assignments they missed upon their return to class, including:

- Obtain notes and assignments from a reliable classmate.
- Check the Makeup Work Calendar for assignments they missed.
- Obtain worksheets from the makeup tray as they enter the classroom.
- See you with questions they have about making up their work. However, they should see you during the times you have set aside for questions regarding makeup work. Students should not ask you to explain missed homework assignments during class.

○ Provide a tray in which students may place their makeup work.

○ Instruct students to see you after class to arrange a time to make up tests and quizzes or to get extra help.

No matter how clear your policy for making up missed work, unfortunately there will be students who will put little effort into completing the work they missed. You may find it necessary to contact these students' parents or guardians, who will likely request a list of missed assignments. For these students and parents or guardians, the following reproducible, "Math Makeup Work," is useful. Noting the missing assignments on the form and sending it home provides a record of assignments that must be completed. Requiring that this sheet, signed by a parent or guardian, be turned in with the assignments facilitates compliance. Be sure to make a copy of the sheet before sending it home.

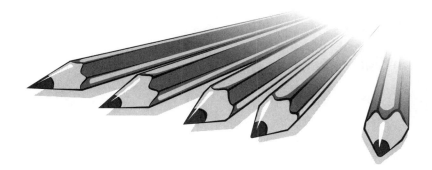

Math Makeup Work

Student: _____   Date _____   Period _____

Course: _____   Date Assignments Are Due: _____

Assignments to be completed and handed in:

_____

_____

_____

_____

_____

_____

_____

_____

_____

_____

_____

_____

_____

Teacher Signature: _____

Student Signature: _____

Parent or Guardian Signature: _____

Many electronic grade books have the capability to generate reports of missing work. If you are using this type of grade book, be sure to keep your assignments current so that you can use the grade book to create an accurate report of assignments.

At the beginning of the year, inform your students and their parents or guardians of your makeup policy. Posting your makeup procedures on a corner of a bulletin board in class, providing a handout to students to keep in their binders, posting your policy on your school's Web site, and explaining it to parents and guardians on back-to-school night emphasizes the importance of making up missed work and minimizes confusion.

To be effective, procedures for makeup work must be clear, practical, and enforced. When you require students to make up work they missed, you are ensuring that they keep pace with the rest of the class and are developing sound work habits. When students make up missed work, they continue to progress in class and experience the satisfaction that comes with accepting responsibility.

PROCEDURES FOR ENDING CLASS

With so much material to cover and so many concepts and skills to teach, most math teachers feel pressure to complete all they plan to do in a period. Sometimes their instruction continues right up to the bell, resulting in a hectic end in which students are itching to rise from their seats even though their teacher is still speaking. Such endings cause confusion, distract students from copying their homework assignment, and result in papers, books, and belongings being left behind.

To avoid a disorganized conclusion of your classes, you must establish routines that keep students on task and enable you to end class in an orderly fashion. Effective concluding routines are a part of a productive learning environment.

Here are some tips for ending your math classes positively:

- Plan your lessons so that you have about three minutes at the end of class to prepare students to leave. Of course, the amount of time depends on your class and the materials to be collected and put away.
- Have student volunteers collect equipment and materials, such as calculators, protractors, and markers, that have been handed out to all students.
- Instruct students to log off computers and return the screens to the desktop.
- Remind individual students to return any materials or equipment they used during class.
- Assign homework and make sure that students write the assignment in their agendas or assignment pads. Circulate around the room if necessary. Provide any special instructions.

- Have a volunteer read the assignment and repeat any instructions.
- Ask if there are any questions about the assignment.
- Instruct students to look about their desks and work areas and pick up stray papers or debris on the floor. They may discard any trash on the way out of class.
- Instruct students to straighten desks and put furniture back where it was when they entered class.
- Remind students to pack up their papers and take all of their books and belongings.
- Dismiss students at the bell. No one should leave his or her seat before the bell and your dismissal.
- Students should leave the room in an orderly manner.

Establishing a routine for ending your math class prevents the last few minutes of the period from slipping into an unproductive time during which students socialize or simply wait for the bell to ring. The routine emphasizes the importance of your class and ensures that students use the entire class period in a productive manner.

Helping Your Students Learn to Follow Directions

Although an ability to follow directions is an important skill in every class, it assumes an even greater urgency in mathematics, a subject in which rules and processes must be strictly followed in order to solve problems. Confusing or ignoring the steps in the Order of Operations, for example, results in incorrect answers. Students who are unable or unwilling to follow directions are less likely to excel in class than students who do follow directions.

There are several steps you can take to help your students learn to follow directions, both verbal and written, including:

- Make certain that the directions you give are clear.
- Establish eye contact with students when you give directions.
- Emphasize important steps. Repeat and clarify as needed.
- Write instructions or directions on the board or project them on a screen. Highlight important words, phrases, and statements with markers, colored chalk, or colored text on computers.
- Insist that students copy assignments word for word.
- To reinforce verbal directions, ask students to repeat them.
- Modulate your voice when giving directions. Raising and lowering your voice helps keep students attuned to your words.

- Explain and model classroom routines and procedures.
- Monitor students as they follow directions. Remind them of directions for routines and procedures as often as necessary.
- Place posters throughout your room to reinforce directions.

The ability to follow directions extends well beyond the math classroom. When your students learn how to follow directions, they learn a life skill that will benefit them many times over in the years to come.

Quick Review for Managing Your Classroom

Managing your classroom effectively is one of your most important and demanding responsibilities as a teacher. A classroom lacking sound management is not as productive as it could be. The following steps summarize how you can manage your math classroom with confidence and proficiency.

- Establish practical classroom routines for managing attendance taking, tardiness, requests to leave the classroom, and distributing and collecting materials.
- Implement procedures for calculator use in class, including distributing, collecting, and maintaining your inventory of calculators.
- Implement procedures for students using computers in class.
- Develop plans to ensure smooth traffic flow in your classroom.
- Promote productive work habits via the following:
  - Starting class efficiently with a math do-now
  - Requiring students to use agendas or assignment pads for writing down homework
  - Requiring students to maintain math notebooks
  - Providing clear guidelines for completing classwork and homework
  - Providing guidelines for cooperative group work
  - Providing guidelines for math journals
  - Developing practical procedures for makeup work
  - Planning effective procedures for ending class
- Help students learn how to follow directions.

Successfully managed math classes are a result of planning and implementation. When you establish effective routines and procedures that support instruction and learning, you create an environment conducive to student achievement.

Building a Positive Environment for Learning Math

Students perform best in classrooms that are inviting, attractive, and safe. An orderly math class—where it is believed that all students, regardless of gender, ethnicity, or background, can learn math; where respect for others is a priority; and where academic success for all is encouraged—becomes a positive learning environment that enables students to realize their greatest potential.

You can create this kind of environment in your math class. By properly arranging the physical classroom, fostering an atmosphere of courtesy and acceptance, and addressing the needs of diverse learners, you can make your class a center for learning math. By building a positive learning environment, you show your students that you value them as members of your class, and invite them to assume membership in a mathematics community.

The Physical Classroom

Building a positive learning environment for math begins with the physical classroom. Whether you teach in a new classroom with all of the latest technology, or a stuffy room in the basement of a hundred-year-old school, there is much you can do to make your students' physical environment nurturing and appealing.

Following are some suggestions:

◎ Make your classroom bright and cheerful. Perhaps you can put in a work order requesting that your classroom be repainted during the summer. If you can choose your colors, consider neutrals, such as soft blues, greens, or light creams. Bright reds, oranges, and yellow can be distracting. Brilliant white can be tiring on students' eyes and may cause glare.

- Promptly place work orders for repairs. Cracked plaster, holes in walls, and broken floor tiles communicate to your students that adults do not care much about their math class. If students conclude that you do not care, they will not care either.

- Make certain that the lighting in your classroom is appropriate. Good lighting promotes learning. Poor lighting is hard on students' eyes and can be distracting. Quality lighting has brightness that is suitable, balanced, and uniform. There should be no glare. If there is, discuss the situation with your principal. Maybe you can order new shades or different lights.

- Arrange furniture to facilitate learning. All students should be able to see the board and projection screen—and you as you teach. Desks, tables, carts, and bookshelves should be arranged in a way that makes it easy for you and your students to move around the classroom. Aisles should always be free of books, gym bags, and papers. (See "Arranging Furniture to Enhance Math Learning" in Section Two and "Achieving a Smooth Flow of Classroom Traffic" in Section Seven.)

- Materials, supplies, and equipment should be easily accessible and stored when not in use. Items should always be returned to their proper places at the end of class. (See "Distributing and Collecting Materials" in Section Seven.)

- Make your wall space attractive and supportive of learning. Display posters that enable students to visualize math concepts—for example, geometry in nature, fractals, and graphs of functions. Post charts that remind students of mathematical processes or facts—fraction, decimal, and percent equivalencies; place value; or problem-solving strategies. (See "Sources for Math Materials and Manipulatives" in Section Three.)

- Display your students' work to help build pride, promote the spirit of learning, and celebrate achievement in math.

- Use bulletin boards to share news about mathematics, math contests, and awards, such as the math student of the month. Consider creating bulletin boards that support the math topics your students are studying. You should also devote space to highlight timely information, classroom rules, a calendar of work assigned, and the bell schedule.

- Display models that illustrate mathematical concepts, such as models of two- and three-dimensional figures.

- Maintain general cleanliness in your classroom. Although your school has a janitorial crew that cleans classrooms at night, having some basic cleaners on hand allows you to clean desks of stray pencil marks or walls of graffiti. Disposable wipes or a bottle of cleaning spray and a roll of paper towels is usually enough. Because some students may have allergies to cleaners, always open the windows when you clean. Quick cleanups can be done between classes, but you should tackle the bigger jobs after school. To clean the dust from the areas surrounding your computers, consider using a small, handheld vacuum.

⊙ Encourage students to help you keep the classroom clean. If a student notices writing on her desk when she enters the classroom, she should tell you so that you can address the problem with the student who used the desk during the preceding period. Instruct students to put materials and equipment back in their proper places when they are done using them, and to pick up papers and any other items that have fallen on the floor. This is especially important before the end of class.

Students spend a substantial amount of time in their math classroom. Making the classroom as appealing and safe as you can will instill in them a sense of pride in the class and in their accomplishments.

Setting a Tone of Respect and Courtesy

Respect and courtesy are prime components of classrooms where students feel safe and accepted. Students who feel comfortable in a class are more open to learning than those who feel uneasy or threatened. They know that their ideas and opinions will be accepted by you and their peers. Because they will not be laughed at or ridiculed if they make mistakes, they will take risks in exploring new ideas and testing conjectures. They will feel confident in assuming an active role in the class.

You can foster a tone of respect and courtesy in your math classes by doing the following:

⊙ Model appropriate behavior for your students. Some of your students may come from homes where respect and courtesy are lacking, and they may be unfamiliar with positive interactions.

⊙ Address your students by name.

⊙ Avoid speaking with sarcasm. Never demean students or make them feel that they or their contributions to class are inadequate.

⊙ Establish and enforce rules that clearly indicate appropriate behavior. Make sure that students understand the consequences for breaking the rules.

⊙ Stop disrespect and discourtesy immediately. Refusing to tolerate hurtful acts on the part of any student sends a clear message that students should always show respect not only to you but also to each other.

⊙ Offer praise to reinforce positive behavior.

- Make "please" and "thank you" a part of your classroom by using these terms yourself.

- Provide opportunities and guidelines for cooperative learning. When students work together, they experience a sense of shared purpose and learn more about each other.

- Foster a sense of teamwork in which you and your students work together to study math.

As a teacher, you are responsible for setting the tone in your classroom; however, you cannot accomplish this alone. Your students must learn what constitutes appropriate social skills and be willing to use these skills in your class.

To help your students become familiar with the behavioral skills that will result in a classroom that is characterized by respect and courtesy, distribute copies of the following "Respect and Courtesy Guidelines." Encourage your students to use these skills in their math class and other classes as well.

Most students view their teachers as role models and pattern much of their own behavior in class after that of their teachers. When they see you treat others with respect and courtesy, they are more likely to act in a respectful and courteous manner.

Respect and Courtesy Guidelines

Name _____ Date _____ Period _____

To be treated with respect and courtesy by others, you must treat others with respect and courtesy. Following the guidelines below will help make you a respected member of your math class. You will feel good about yourself, the class, and your achievements.

1. Be polite to everyone.

2. Follow school and classroom rules.

3. Listen to others when they are speaking. Everyone feels that his or her ideas are important, just as you feel your ideas are important.

4. Do not talk when others are speaking. Raise your hand to speak in class. Wait to be called on.

5. Express your ideas clearly and try to understand the ideas of others.

6. Contribute to class activities and discussions.

7. Be willing to do your part in group work.

8. Share materials and supplies.

9. Do not disturb others from their work.

10. Never bully, threaten, or make fun of others.

11. Accept individual differences. Be tolerant of others.

12. Be honest.

13. Do not cheat.

14. Remember that "please" and "thank you" are little words that mean a lot. Use them often.

15. Celebrate your success and the success of your classmates.

Preventing and Responding to Bullying

Bullying is a serious problem in schools across the country. By making victims feel threatened and powerless, a bully not only batters the victim's self-esteem but undermines his focus on schoolwork.

Bullying assumes many forms. It can be physical or verbal. Physical bullying can result in violence and bodily harm, and although verbal bullying is not violent, it too can be extremely hurtful with name calling, racial slurs, teasing, or ridiculing. Bullying may be direct—a face-to-face confrontation—or indirect—a whispered threat while passing in the hall. One student may bully another; or several students may bully one victim.

While anyone can be a victim of bullying, most victims are usually younger, smaller, or perceived by the bully as being weak. Victims may be "different" from the majority of students who attend a school. They might belong to a different ethnic group, come from a different background, or have a different economic status. They may be isolated, lacking the support of a strong network of friends, which makes them easier targets. They may simply be a random target that the bully finds he can coerce.

The impact of bullying is significant. Because the victim no longer feels safe at school, his schoolwork suffers. To avoid the bully, he may feign illness to stay home, or may become ill from worry. The impact extends beyond the victim. Other students in the class may feel powerless and confused when this behavior is permitted, and they may come to feel threatened even though they are not victims. School is no longer a place of learning but a place to be feared.

Despite its severity, bullying can be difficult for teachers to recognize. It is often covert and may appear to be harmless horseplay, or dismissed as a part of growing up. Compounding matters, victims may deny that they are being harassed. Bullying in any form or degree, however, should never be tolerated.

The following steps can help you prevent and respond to bullying in your classroom:

- Encourage your students to treat each other with courtesy and respect.
- Make it clear to everyone that bullying will not be tolerated.
- Understand that you are expected to address bullying. Find out about your school's policy and your responsibilities regarding bullying. Follow the policies closely.
- Be alert to the signs of bullying. Although many signs of bullying are often a result of other conditions, a student who exhibits the following may be the victim of a bully:
  - Repeated target of teasing, horseplay, or ridicule
  - Falling grades for no other apparent reason

- Decrease in ability to concentrate
- Increase in absence, particularly if prior attendance was good
- Frequent complaints of headaches, stomachaches, and a need to see the nurse
- Withdrawal from friends
- Mood changes—increasing sadness, anxiety, moodiness
- Decrease in self-esteem
- Inexplicable bruises, cuts, and other injuries
- Loss of possessions, particularly money
- Refusal to talk about problems

- Always take prompt action once you suspect bullying. For example, if you overhear a student threaten another, address the issue of bullying immediately. Do not simply tell the students to focus on their work with the intention that you will speak to them about the incident later. You should never delay action on your part regarding bullying.

- Be considerate of the feelings of victims, who often feel afraid, embarrassed, and powerless.

- Notify administrators and guidance personnel when addressing an incident of bullying. They may be able to provide some insight into the behavior of both the bully and the victim.

In addition to addressing bullying in your classroom, you may also provide the following strategies for your students to help them guard against being bullied:

- Students should try to avoid bullies. If they know that a bully leaves school through a particular door, they should leave through another.

- Students should use the buddy system. Friends can provide support and keep bullies away.

- When facing a bully, students should try to keep their emotions in check. Bullies enjoy making others feel afraid and powerless.

- Students should act as if they do not care what the bully says. Most bullies lose interest in those they cannot upset.

- Students should tell an adult. Parents or guardians, teachers, guidance counselors, or the school's principal can all help.

Despite all of your efforts to establish a positive tone in your classroom, you may have to address a bully and a victim. Acting promptly and decisively as soon as you become aware of the problem can usually end it. When students understand that you will not tolerate bullying, they will feel safe in your classroom and be able to concentrate on learning math.

Creating a Positive Math Environment for Diverse Students

Each day you no doubt work with students of various abilities, personalities, ethnicities, and family backgrounds. In the course of a day, you will likely work with gifted students, underachieving students, mainstreamed special-needs students, students with attention deficit disorder, students with 504 plans, at-risk students, economically disadvantaged students, and students for whom English is a second language. Sprinkled among these students will be average students who, because they do not call attention to themselves, blend in with the class, but may have unique needs of their own.

All of your students have needs that will affect their performance in your math class. Your success in addressing their needs is a major factor of their success in learning math. It is also a major factor in your school year being a satisfying and productive one.

GIFTED MATH STUDENTS

Gifted students who have a high aptitude for mathematics can be both enjoyable and challenging to teach. Because these students learn skills and concepts in math quickly, they can easily become bored with topics they have mastered and frustrated if they feel the class is not moving ahead fast enough. They require a variety of approaches, depending on their readiness and interests, and a curriculum that is deeper and more extensive than that of a typical math class. By modifying the content of your lessons and adapting your instruction to meet their needs, you will ensure that advanced students will remain interested in your class and experience a successful year.

Consider the following strategies to meet the needs of your students with superior math abilities:

- Set the pace of your instruction according to the capabilities of your students. The pace of a class of students who have high mathematical ability should be faster than the pace of an average class in the same subject. Even in a class of students with mixed abilities, you can modify instruction for your gifted math students.

- Gifted students usually are enthusiastic learners. Stimulate their interest by offering plenty of opportunities to forge ahead. Independent work, explorations, and challenging activities such as math puzzles and games offer excellent ways to provide students with material that goes beyond your standard lessons. You may also wish to give these students the chance to teach or tutor other students.

- Assign projects in which students are required to research, analyze, and organize data, make conjectures, and draw conclusions. Projects, which may be done individually or in groups, allow advanced students the freedom, within a structured activity, to make full use of their abilities. Sharing the results of projects can lead to discussions that examine problem-solving strategies and extend mathematical concepts into other subject areas.

⊙ Plan lessons around higher-level skills. Advanced students quickly master the basics and benefit enormously from work that involves critical thinking. Particularly consider the following:

- Present interesting problems that arise from real-life situations.

- Ask students to create and solve problems of their own.

- Require students to provide detailed answers, oral and in writing, when solving problems.

- Assign problems that have several solutions.

- Avoid repetition.

- Provide activities that require students to approach problems from various perspectives.

- Allow students to use math manipulatives to explore concepts.

- Assign enrichment or extension activities. These may be found in your text, resource books, or online.

⊙ Include technology in your lessons whenever possible. Focus on showing your students how technology—particularly computers and calculators—can support the study of mathematics. Calculators can be used as a tool to explore and solve real-life problems. The Internet is a source of problems, resources, and virtual manipulatives that may not be included in your text. Mathematics software can be used to make and test conjectures. (See "Technology" in Section Three.)

⊙ Consider placing gifted students with other gifted students when working in groups. Although it is often desirable for students to work with students of varying abilities and interests in groups, it is just as important for advanced students to work with other advanced students. This situation provides high-ability students the chance to experience the enthusiasm and ideas of their peers.

⊙ Encourage your students to compete in math contests. Consider Math Counts, grades 6–8 (www.mathcounts.org). Also consider contests sponsored by the Mathematical Association of America, including American Mathematics Contest 8, grades 6–8; American Mathematics Contest 10, grades 9–10; and American Mathematics Contest 12, grades 11–12 (www.maa.org/subpage_6.html).

Providing instruction for gifted or advanced math students is much like running a race where you stay a half-step ahead of everyone else. You dare not miss a step because your students will then surge ahead of you. To keep that step ahead, visit the Web site of the National Association for Gifted Children (www.nagc.org). You may also search the Internet with the term "teaching gifted math students" to locate many other helpful Web sites.

UNDERACHIEVING STUDENTS

An underachieving student is one who does not work up to his or her potential. Some students may underachieve in many classes and others may be underachievers in one or two.

There can be many reasons for a student's underachievement. She may be afraid of failing to meet the high expectations of her parents; she may be involved in so many outside activities that schoolwork, and particularly math, is not a priority; or she may be unmotivated because of lack of parental support, disinterest in math, or problems at home or at school. Sometimes a student may underachieve because she is not challenged in her schoolwork. Whatever the reason, underachievement hinders a student's progress in learning math and realizing her full potential. Underachieving students present you with the challenge of helping them even though they may not be willing to help themselves.

Following are strategies you can use to work effectively with underachieving students:

- Express to your students your expectations that everyone can do well in your math class. Many underachieving students convince themselves that they cannot learn math. They may make excuses by saying "I never could do math," or "Math is my worst subject."

- Make your instruction and activities as interesting as you can. Underachieving students frequently lack motivation. Comments such as "Why do we need to learn this?" and "I'll never use this" are indicators of disinterest. Relate your math lessons to the interests of your students whenever possible. For example, when teaching percents and discounts, using a problem that shows how to find the cost of a CD discounted 20% is more meaningful to students than simply computing the percents of numbers.

- Encourage all your students, particularly your underachievers. When students feel that you believe they can do well in your class, they will work harder to complete assignments on time. If they feel that you believe in them, they will believe in themselves.

- Review the results of standardized tests for information about underachieving students' general abilities in math. Supplement any weaknesses they may have.

- Because many underachievers find the typical tasks of school work tiresome, teach your students study and organizational skills to help them become efficient learners. Especially note the importance of the following:

  - Bringing all necessary materials to math class, including texts, math notebooks, agendas or assignment pads, and pencils
  - Staying on task and completing classwork
  - Writing down assignments in their agendas or assignment pads
  - Organizing their math notebooks
  - Managing their time and prioritizing their tasks
  - Starting long-term assignments, such as math projects, well in advance of due dates
  - Breaking long-term assignments into manageable parts and deadlines

- When underachieving students have difficulty with assignments, offer support rather than criticism. Ask them to try to complete all of the work. If they are unable to finish, ask them to write down the reason why the problem, or problems, proved to be difficult for them. This will make it easier for you to identify and clarify any confusion.

- Offer extra help for your students during and outside of class. Particularly encourage underachieving students to come after school when they need help with their assignments and preparing for tests.

- Be liberal, but genuine, with praise when underachieving students complete their work on time and have done a nice job.

- Work with guidance counselors and the parents or guardians of underachieving students to help the students acquire good work habits and complete their work on time. You may gain some insight about a student's situation of which you may not have been aware.

- If you feel it is necessary, refer the student to the child study team for evaluation.

You can find much information on the Internet for working effectively with underachieving students. Search with the term "teaching underachieving math students."

Everyone likes to experience success, even students who are chronic underachievers. Strive to help all your students reach their potential.

MAINSTREAMED SPECIAL-NEEDS STUDENTS

Students with special needs are mainstreamed into regular classrooms whenever possible so that they have the opportunity to interact with the general student population. You will undoubtedly have some special-needs students in your classes. It is likely that you will work with special education teachers and para-educators to provide a positive learning program for some, if not all, of the special-needs students in your class. (See "Working with Other Teachers" and "Working with Para-Educators" in Section Four.)

The term "special needs" is a broad one that includes numerous disabilities, limitations, and weaknesses that, if left unaddressed, may impair academic or social success. In some cases the conditions will be severe and require significant accommodations on your part; in others, they may be mild and require little modification of your daily lessons, plans, and routines. You must be aware of the needs of all your mainstreamed students and the modifications you must make for them.

The following strategies can help you provide a positive educational experience for the special-needs students in your classroom:

- Be fully cognizant of your students' individual needs. Check the records of your students and consult with their special education teachers and, if necessary, the members of the child study team.

- Read, implement, and follow the IEP (Individualized Education Program) of each student. Developed by a student's teachers, parents or guardians, and support staff, an IEP is a written educational program that details a student's educational goals, levels of performance, and required classroom modifications. Remember that IEPs are legal documents that must be followed.

- Refer to the student's IEP throughout the school year to be sure that the modifications you have implemented are consistent with those required by the IEP.

- Modify your expectations and instruction as necessary so that special-needs students can be successful in your class. For example, consider the following:
  - Impress upon your special-needs students (along with your other students) the importance of following the class's rules, routines, and procedures.
  - Shorten the length of assignments to enhance an opportunity for success. For example, instead of requiring a special-needs student to complete the entire assignment of twenty problems, require him to complete fifteen.
  - Divide long assignments into smaller units so that students do not feel overwhelmed.
  - Modify assignments to match the learning styles of special-needs students. (See "Addressing the Needs of Diverse Learners" in Section Ten.)
  - Monitor special-needs students frequently during class to help keep them on task and assess their progress.

- Provide prompt feedback. Quick and consistent feedback assures students that they are doing the work correctly. It also enables you to correct mistakes before the student moves too far along with the assignment.

- Provide clear and concise instructions. Make sure that special-needs students understand exactly what they are supposed to do to complete an assignment.

- Write clearly on boards or transparencies for an overhead projector. Use different colors to highlight key words, and leave space between your notes to avoid crowding information.

- Provide activities for special-needs students to work cooperatively with other members of the class.

- Adapt assessment methods. For example, if a student has a severe weakness in reading, his special education teacher might read the word problems on a test for him. The student can then complete the problems, which will demonstrate his math competence.

○ Make the necessary accommodations to support the learning experience of special-needs students. For example, if a student is to receive extra time for completing tests, be certain that you provide the necessary time. Consulting a student's special education teacher regularly can ensure that you are providing the appropriate modifications.

○ If a special education teacher is assigned to your classroom to help you meet the needs of mainstreamed students, work closely with her to build a positive learning environment and to plan and deliver effective instruction. Though you should be responsible for planning activities that meet the course's requirements of all the students in the class, the special education teacher should accept responsibility for adapting the material and assessments to meet the needs and learning styles of special-needs students.

○ Also work closely with support personnel such as para-educators, guidance counselors, social workers, and school psychologists. Teamwork is essential for providing a comfortable and effective classroom environment for special-needs students.

○ Keep in mind that some disabilities are not readily apparent. Some mainstreamed students may appear to be quite average, especially if they want to fit in with the class. Avoid making the mistake that these students simply need to work harder to learn math. Their disabilities must be addressed in planning, instruction, and assessment.

○ Be supportive of the mainstreamed students in your classes. Encourage them and provide them with the extra help they require to succeed. Applaud their efforts.

○ Work closely with the parents and guardians of your mainstreamed students. Keep them informed of their child's progress, and of upcoming tests, projects, and special activities in class.

You can find much useful information on the Internet regarding special-needs students. Two of the most helpful Web sites are LD Online (www.ldonline.org) and Learning Disabilities Association of America (www.ldaamerica.org). Other valuable Web sites can be found using the search term "teaching math to special-needs students." With the proper nurturing and instruction, your special-needs students can help to make your school year extremely satisfying.

STUDENTS WITH 504 PLANS

A 504 plan arises from Section 504 of the Rehabilitation Act that guarantees that individuals with disabilities do not suffer discrimination in federally funded programs or activities. Section 504 of the Act addresses education, and states that schools cannot discriminate against students with disabilities and requires schools to make accommodations to ensure that discrimination in the learning environment does not occur. A 504 plan for a student outlines specific classroom and instructional modifications that teachers must provide.

504 plans are designed to assist students with special needs who are placed in a regular classroom. These students do not require instruction from a special education teacher, yet they need adjustments in the classroom setting to enable them to perform at the same levels as their classmates. A variety of disabilities are covered in 504 plans, some of the most common being:

○ Attention deficit disorder (ADD)
○ Attention deficit hyperactivity disorder (ADHD)
○ Impairments of vision or hearing
○ Communicable diseases
○ Recovery from chemical dependence
○ Chronic illnesses, such as diabetes
○ Chronic conditions, such as asthma or severe allergies
○ Confinement to a wheelchair

The 504 plan details the accommodations that students must receive. Of course, accommodations vary according to an individual student's disabilities, but may include the following:

○ Assignments or testing conditions may be adjusted. For example, additional time may be provided, or test questions may be modified.

- A student may be required to sit in the front row so that she can hear more clearly.
- A student may be allowed to go to the school nurse during class for administration of medication.
- A student suffering from diabetes may be permitted to eat in the classroom.
- A student may be provided with a set of books for home use.
- A student may be provided with copies of class notes.
- The classroom may need to be wheelchair accessible.
- Regular updates may need to be made to parents.

At the beginning of the school term, you should receive copies of 504 plans for your students. You should then meet with the 504 plan administrator in your school to discuss each student and clarify your responsibilities in making the necessary accommodations.

As you are required by law to make these accommodations, you should keep documentation for each student showing that you have implemented his or her plan, or dutifully attempted to implement the plan. Maintaining a file of student work (especially adjusted math tests and assignments), personal notes, and copies of correspondence to parents or guardians and administrators will detail your efforts. If a plan is not working, inform your principal and 504 administrator as well as support staff, such as guidance counselors, the school nurse, and school psychologist. Also inform parents or guardians. A review meeting should be arranged during which the overall goals and effectiveness of the original plan should be discussed and any necessary adjustments should be made.

To find more information about 504 plans, visit the Web site of the U.S. Department of Education (www.ed.gov) or your state department of education and search for "504 plan." A general search with the same term will lead to many helpful Web sites.

The accommodations required by most 504 plans modify the learning environment sufficiently to meet the needs of the students. In most cases, small modifications produce major gains in learning.

STUDENTS WITH ATTENTION DEFICIT DISORDERS

Students with attention deficit disorder (ADD) or attention deficit hyperactivity disorder (ADHD) can be among your most challenging. These students typically have trouble concentrating for long periods, lack organizational skills, are easily distracted, and often shift from one task to another before completion. They may also exhibit impulsiveness, inattentiveness, and difficulty with social interactions. Whereas the most significant characteristic of a student with ADD is inattentiveness, the most significant characteristics of a student with ADHD is inattentiveness coupled with hyperactivity. For these students to achieve academic success, you must address their individual needs with consideration and consistency.

The following suggestions can help you to develop a productive learning environment for your students with attention deficit disorders:

- Learn as much as you can about these students. Depending on the individual, some ADD and ADHD students may have IEPs, while others have 504 plans.

- If the student is mainstreamed, consult frequently with his special education teacher, who will be able to offer insights to the student's specific needs. Special education teachers can also suggest approaches that have worked for the student in the past.

- Enlist the support of parents and guardians of ADD and ADHD students. They can provide substantial help. Ask them to work with their children at home to ensure that they complete homework and are organized for school the next day.

- Work with the students to improve their study and organizational skills such as coming to class prepared and on time, taking notes, and writing down assignments in agendas or assignment pads. Realize that you will need to remind ADD and ADHD students of such tasks often.

- Have extra pencils and paper readily available, and keep your classroom organized and free of clutter. Instruct students to clear their desks of items and materials that they will not use during class. Minimizing distractions helps students to remain focused.

- Establish consistent classroom routines. ADD and ADHD students work best in structured environments.

- Post rules for the classroom's routines and procedures on a bulletin board where students can see them. Make the display colorful and attractive; reinforce rules regularly and consistently.

- Seat ADD and ADHD students close to you so that you can monitor them easily. Consider seating them next to students who are good workers. Not only will ADD and ADHD students be less distracted, they will be near positive role models. Avoid seating them near students who might disturb them, as well as near doors, windows, pencil sharpeners, waste baskets, and other areas of high traffic or interest.

- Give clear directions, one step at a time. Emphasize the key points of any instructions. Have a student volunteer repeat directions.

- Give shorter but more frequent assignments. If necessary, allow extra time for completing assignments.

- Change tasks frequently. Do not stay on one topic too long.

- Check to make sure that your ADD and ADHD students have written down their homework assignment correctly.

- Involve students in the classroom's activities to keep their attention focused. Move around the classroom to keep students on task. Offer encouragement for their work and contributions.

- Help students to channel their energy in appropriate manners. Ask them to write sample problems on the board, and hand out and collect papers and materials.

- Monitor the classwork of your students closely. Check their work after they have completed four or five problems and give them immediate feedback. Small successful tasks will encourage them to move forward.

- Allow your ADD and ADHD students to use calculators for computation. This frees them from worrying about making mistakes in basic operations and permits them to focus on completing their work.

- Use manipulatives, number lines, and graph paper to visually demonstrate math concepts and relationships.

- Use diagrams, charts, and graphic organizers to help students clarify their work and thinking.

- Encourage students to reread word problems before trying to solve them. This will enhance their understanding of what the problem is asking. Instruct them to underline key information and words that indicate the operation(s) they should use to solve the problem. Remind them to double-check their work.

- Post lists of steps for solving word problems and strategies for problem solving on the bulletin board.

- Provide a sheet of formulas, which eliminates the need for memorization and permits ADD and ADHD students to concentrate on solving problems.

- Review frequently, highlighting important points so that students have time to fully reflect on and absorb the material.

- Use computer games to help students practice basic math skills. Web sites such as www.funbrain.com offer an assortment of games. Encourage your students to play the games at home.

- Use peer tutoring to reinforce skills for ADD and ADHD students and provide more practice with math. Set up partners for activities whenever possible.

- Remember to acknowledge and praise positive behavior and effort.

To find more information about students with attention deficit disorders visit the Web site of Children and Adults with Attention Deficit/Hyperactivity Disorder (www.chadd.org). A search using the term "math students with ADD and ADHD" will yield many helpful Web sites.

Students with ADD or ADHD do best in classrooms in which their teachers understand their needs and provide a supportive learning environment, yet hold high expectations for their academic success. Fully engaging these students in daily activities will help them to become valued members of the class.

STUDENTS WHO ARE AT RISK OF DROPPING OUT OF SCHOOL

Each year some 25 percent of U.S. high school students drop out of school before graduating. In some inner city schools the drop-out rate is closer to 40 percent. Students who may drop out of school are known as at-risk students.

At-risk students may drop out for numerous reasons, including:

- Poor academic skills and little, if any, academic success
- Repeated failure
- Undiagnosed or misdiagnosed learning problems
- Family problems
- Lack of parental supervision and support
- Social and peer problems
- Language barriers
- Problems with substance abuse
- Emotional problems
- Pregnancy
- Chronic illness

At-risk students typically experience little academic success and suffer low self-esteem. They tend not to participate in school activities, have minimal interest in school, and see little value in what they might learn in class. For them, school is a negative place.

Along with providing at-risk students the academic and emotional support they need, the key to keeping them in school is to make school a positive environment. An extra effort on your part can be the difference between an at-risk student leaving or staying and graduating.

Following are some strategies that you should consider to help at-risk students in your math class:

- Let at-risk students know that you care about their academic, personal, and future success. Knowing that someone truly cares can be a major incentive to remain in school.

- Help your students identify future goals and show them how education can help them achieve these goals.

- Be attentive to the needs of at-risk students. Your consideration of them as individuals may be more meaningful to them than you can ever imagine.

- Work with other professionals to secure any special help that the students may need, for example, remediation, child care services, medical care, substance abuse awareness programs, and bilingual instruction.

- Assign meaningful work that has a clear value for students. When students see the benefits of learning, they are more likely to stay in school. Try to relate math to their lives and focus on practical math skills. For example, you might consider a project in which students search the classified ads (in newspapers or online) for a job. They should research the qualifications required, hours, and wages. Ask your students to calculate their weekly earnings and anticipated transportation costs in traveling to and from work. You might also discuss payroll deductions and ask students to estimate what their net pay would be.

- Use technology whenever possible. Allowing students to use calculators enables them to solve problems that might otherwise frustrate them. Calculators also provide immediate feedback.

- Involve at-risk students in class and group activities. Because they are contemplating dropping out, these students often stay in the background and do not participate. By encouraging them to be a part of the class, you give them a chance to connect with other students.

- Provide at-risk students with opportunities in which they can be successful in class. Maybe an at-risk student who is a great guitar player can explain how musical notes are based on fractions. (An octave, for example, is a tone that is eight full tones higher or lower than a given tone. Octaves are based on eighths.) The student may even wish to play a few notes to demonstrate this.

- Be sure to provide activities that address at-risk students' learning styles. Although you should do this for all students as much as possible, it is particularly important for at-risk students. Small, consistent successes can be a reason to stay in school. (See "Meeting the Needs of Diverse Learners Through Instruction" in Section Eleven.)

- Make certain your at-risk students know that you are available for extra help. Make them feel welcome when they come for help.

- Work closely with the parents and guardians of your at-risk students, as well as with support personnel, to provide these students with the reasons and incentives to remain in school.

○ Explore options for at-risk students. Perhaps there is a co-op program in which students spend the morning in school and work part-time in the afternoon.

You can learn much more about how to address the needs of at-risk students in your class by searching the Internet with the terms "at-risk students" and "teaching math to at-risk students." These students have the most to lose and very little, if anything, to gain by dropping out of school. You and the other professionals who work with them need to provide them with a positive and rewarding environment that will encourage them to stay in school and graduate.

ECONOMICALLY DISADVANTAGED STUDENTS

Students whose families live in poverty often struggle in school. They do not enjoy the advantages and experiences of their peers who come from typical middle class families. Students from poor families may lack money to purchase basic school supplies, a calculator for math class, or new clothes for school. Certainly they cannot afford many of the items that their classmates take for granted, and they often drop out of school to earn money for themselves or to help their families. For many, school is simply not a priority.

The following suggestions can help you to make school, and particularly your math class, a place that economically disadvantaged students will appreciate:

○ Provide meaningful math lessons that tie into real-life skills. Include word problems to which students can relate, and emphasize the value of math for future careers.

○ When planning activities, keep in mind that economically disadvantaged students will not have had the same experiences as other students. For example, it is unlikely that students from poor families have gone on many family vacations, attended major sporting events or concerts, or traveled to another country. Avoid making repeated references to situations and conditions with which they cannot identify.

○ Maintain high expectations for all your students. Express your belief that everyone in your class—no matter his or her background—can succeed through hard work.

○ Provide access to technology, especially calculators and computers, that economically disadvantaged students might not have at home. Perhaps your school will allow you to loan a calculator to a needy student for the entire school year.

○ Provide basic supplies and materials as necessary in case students run out of their own.

○ Be especially sensitive to the feelings and needs of economically disadvantaged students. Avoid saying things that might make them feel uncomfortable, for example, casually discussing the latest fashions which they may not be able to afford. If you notice students teasing or ridiculing their disadvantaged classmates, promptly intervene and correct the behavior.

Most schools have students who are economically disadvantaged. Although these students might not enjoy the economic advantages of other students, they can one day be just as successful. The road to their success starts with school.

STUDENTS WHO SPEAK LITTLE OR NO ENGLISH

Imagine yourself as a student in another country. You understand only a smattering of the language; you are trying to adapt to a culture that is vastly different from your own; and you are going to a strange school where you can barely communicate. To describe the feelings of such students as being overwhelmed is without question an understatement.

As America grows increasingly diverse, so do its classrooms. Many teachers have students who come from different backgrounds and countries. Many of these students have a limited ability to speak English, or they may not speak English at all. They are often labeled with terms such as ESL students (English as a Second Language), LEP students (Limited English Proficiency), and ELL (English Language Learner). Labels aside, all these students share a common fact: They have difficulty speaking, reading, and understanding English. As the background of each student is different, you must tailor a program that addresses his or her needs in your math class.

Following are some suggestions:

○ If you have a student who speaks limited or no English, consult her ESL teacher. He may be able to provide you with background information on the student, offer strategies you can use in modifying your math program for her, and explain the extent of her ability to speak and understand English. Sometimes a student may not be able to speak English but is able to understand much of what you say.

○ Be patient when students speak to you or answer questions. Some students will formulate their thoughts in their native language and need time to then express those thoughts in English. Showing impatience or hurrying them will only cause frustration for both of you.

○ Ask the ESL teacher to help students with basic math vocabulary they need to know in your class, such as *sum, difference, multiplication, division,* and so on. Regularly provide lists of words to the ESL teacher before you introduce these terms in class. Many math texts today contain a Spanish glossary.

If your text does, be sure to provide a copy to the ESL teacher and utilize it with your Spanish-speaking students.

◎ If another staff member speaks the student's language, ask if this person is willing, when necessary, to serve as a translator.

◎ If possible, partner the student with a dependable student who speaks his language.

◎ Use visual aids and manipulatives to illustrate math concepts. Being able to direct a student's attention to a picture or example of a square when you say "square" is enormously helpful.

◎ Be friendly and open to your students. Greeting them by name as they enter the classroom can help put them at ease. A friendly face is always welcome in a strange place.

◎ Always speak slowly and enunciate your words clearly. Use accurate terms when speaking about math. Whenever possible, match words with symbols, representations, or objects.

◎ Try to spend some extra time explaining the day's work to students. Keep the language basic.

◎ Whenever possible, write directions on the board and say the directions aloud.

◎ Label items in the classroom to help students learn simple words. The door, windows, desks, paper, pencil sharpener, waste basket, and mathematical models are good places to start.

◎ Seat students who speak limited or no English in the front of the class where they can hear you more clearly. This will help them learn English.

◎ Use bilingual dictionaries to help with communication.

◎ Use charts, tables, and graphs to illustrate and display information in a clear and logical manner.

◎ Allow students to use calculators for problem solving.

◎ Include students who speak limited or no English in groups for cooperative activities. Pair them with students who, by nature, are patient and willing to help in instances where the student has difficulty understanding.

◎ Look for ways that students with limited or no English can shine in class. For example, some students who come from other countries may understand the metric system and be familiar with metric units.

◎ Be willing to adjust your goals for students who speak little or no English. Even though their mathematical ability may be sufficient, or superior, they will have trouble reading directions and solving word problems. Note: Before adjusting your program, obtain approval from your supervisor. In many schools, guidelines for grading students who speak limited or no

English are vague. Clearly it is unfair to fail a student who is competent in computation but cannot pass tests and quizzes because she is unable to read English. A reasonable accommodation should be made.

- Avoid setting unreasonable expectations for students.

A search of the Internet using the term "instruction for ESL students" or "students who speak limited English" will provide many helpful Web sites for teachers of students who speak little English. Two that we find particularly useful are Yahoo!'s Babel Fish (http://babelfish.yahoo.com), a free online language translator, and I Love Language (www.ilovelanguage.com), which provides numerous language resources.

Because of the universal properties of numbers, for many students who speak little or no English mathematics is a subject at which they may excel. Helping these students achieve success in math is a major step to helping them become acclimated to your school and community. (See "Working with Parents and Guardians Who Speak Limited English" in Section Fourteen.)

It requires great commitment and patience to meet the needs of all the students in your classes and satisfy your course requirements. You must be willing to adjust your instruction and make accommodations, while at the same time creating and maintaining a positive classroom environment.

Avoiding Gender and Ethnic Bias in Math Class

As a math teacher you have probably heard or read comments such as the following: "Girls don't have the same aptitude for math as boys." "Boys are more competent with technology than girls." "Asian students excel at math."

Of course, many girls do very well in math, many boys are less computer savvy than many girls, and though many Asian students do excel in math, some others do not. Despite such evidence, however, the stereotypes persist.

So pervasive are stereotypes in our society that they can creep into our subconscious and influence our behavior. This can be particularly troubling for teachers, as it can undermine a teacher's efforts to provide an equitable learning environment for her students.

Consider the following to help you avoid gender and ethnic bias in your class:

- Maintain the same high—but realistic—expectations for all your students. Let students know that you believe everyone—girls and boys, regardless of ethnicity—can learn math in your class. Success comes from effort, not from background or gender.

- Make all materials and activities of the class available to all students.

- Praise and support all students in your class.

- Do not designate boys for some duties, for example, moving desks, and girls for others, such as passing out papers.

- Never say that one ethnic group has more aptitude, or is expected to do better in math, than another. Students will believe you and live up to your expectations.

- Call on all students equally to answer questions, regardless of their sex, race, or socioeconomic status.

- Avoid directing the "tough" questions to certain students. Likewise, avoid asking only some students the "easy" questions. Your students will quickly conclude that you believe some of them are more capable than others.

- Provide activities for cooperative learning. Create groups of mixed gender and ethnicity.

- Encourage all of your students to assume leadership roles and contribute to class activities.

- Provide equal opportunities for all students in your class. For example, encourage all students to try out for the school math team and take part in math contests.

- Avoid bias in assignments and activities. A worksheet that contains word problems should show examples of girls, boys, and various ethnicities in different but equal roles.

- Use gender-neutral language. Instead of *chairman*, use *chairperson*; instead of *policeman*, use *police officer*; instead of *councilman*, use *councilperson*.

- Be sensitive to the content and delivery of material as well as your interaction with students. Do not say or do things that students might interpret as a slight because of their gender or ethnicity.

As teachers, we all want all our students to do well. We want them to master the math concepts and skills they will need to attain their future goals. We can best accomplish this challenge through math classes that satisfy curriculum requirements and provide encouragement and support equally to every student.

Appreciating Cultural Diversity

In schools across the country, classrooms are filled with students of different races and cultural backgrounds. In some cases, because of intolerance of differences, the diversity can lead to divisiveness and strife. In others, because teachers, administrators, and support staff set examples of acceptance, diversity becomes a resource in which students learn and benefit from the differences of their peers.

To foster a sense of appreciation and community from the diversity in your classroom, consider presenting your students with math activities that originated in other countries. Some common activities, their lands of origin, and mathematical focus include tangrams (China, geometry), magic squares (China, Egypt, India, computation), and origami (China, Japan, geometry). If your school, like many, has a multicultural day, perhaps you can provide a sampling of math activities from around the world. Searching the Internet by using the term "multicultural math activities" will provide you with many ideas and resources.

In addition to planning multicultural activities, the following suggestions can help you expand your students' understanding and appreciation of each other's talents, backgrounds, and differences:

- Establish and maintain a policy of acceptance of individual differences. Make sure everyone understands that intolerance is unacceptable.
- Treat all students equally.
- Be willing to learn about the customs, traditions, and holidays of different ethnicities in your class.

A reference to a student's homeland evokes a sense of pride, and is a springboard from which students can learn more about their own background as well as the backgrounds of others. Through math games and activities that have their roots in another country, students gain an understanding of math from a fresh cultural perspective. They will realize that mathematics is worldwide in scope and importance.

Helping Students Overcome Math Anxiety

Many students worry that they cannot learn math. Perhaps they find mathematics to be abstract, their parents do not offer enough support, or they have simply not done well in math in the past. Whatever the reason, they feel that math is a difficult subject. For many, these feelings manifest in poor grades.

The thinking of these students is straightforward: "Since I can't learn math, why try?" Consequently they do not work hard and their progress is undermined. This only makes learning new skills harder, which perpetuates the idea that they cannot learn math.

For these students, math class is an uncomfortable place. They dread going to class, and once there are impatient for class to end.

To ease any math anxiety for your students, you must make your class a positive experience for them. You must create a supportive learning environment in which encouragement is abundant, praise is genuine, and mistakes are viewed as steps to finding solutions. By expressing your belief that all students can be successful in your class, you can build your students' confidence and give them the incentive to work hard. To help students overcome math anxiety, consider handing out copies of "Steps to Beat Math Anxiety," which follows.

Most students want to do well in school. Although many factors can adversely affect achievement in your math class, math anxiety should not be one of them.

Steps to Beat Math Anxiety

Name _____ Date _____ Period _____

If you worry a lot about doing well in math class, you may have math anxiety. The following tips can help you overcome your anxiety and do better in math.

1. Know that everyone can learn math if he or she works hard.

2. Think positively. Instead of telling yourself that you cannot do well in math, tell yourself that you can.

3. Participate in class activities.

4. Pay attention to your teacher's lessons and explanations.

5. Take accurate notes during class. Review your notes when you do your homework.

6. Ask questions if you do not understand something.

7. Complete all of your classwork and homework.

8. Work hard. Do not give up easily on challenging problems.

9. When you are working on a tough problem and you start to worry that you cannot solve it, give your mind a short break. Take a deep breath, then exhale deeply. Relax and think of something pleasant—a fun time you had with your family or friends. Return to the problem fresh and ready to find a solution.

10. Find a math study buddy. Working together makes learning math easier.

11. Begin studying for math tests a few days ahead of time. Do not wait until the night before. Preparation is a key to success.

12. Look at math anxiety as being a problem you can solve. You just have to work hard.

Quick Review for Building a Positive Environment for Learning Math

Students who feel good about their math class usually do better than students who feel unwelcome, unsafe, or uncomfortable. A positive classroom environment is the foundation for a successful year for you and your students.

The following tips outline the key elements for creating a positive environment for learning math:

- Make your classroom bright and cheerful for students. Proper lighting, a practical arrangement of furniture, accessible materials, attractive bulletin boards, and overall orderliness and cleanliness can make any classroom appealing.
- Establish a tone of respect and courtesy in your classroom.
- Make sure that the classroom is a safe environment for your students.
- Promptly address bullying or harassment. Take steps to prevent these problems.
- Create a learning environment that addresses the needs of diverse students.
- Avoid gender and ethnic bias in your planning, delivery of instruction, and interaction with your students.
- Be cognizant of and attentive to cultural diversity in your class. Plan and present activities that celebrate the different cultural backgrounds of your students.
- Work with your students to help them overcome anxiety about math. Students who feel confident about themselves and their place in the class are more successful than those who view math class with discomfort and a lack of confidence.

Whereas a positive classroom environment supports learning, a poor one detracts from it. By making your classroom welcoming and appealing to your students, you will provide them with an essential element for achievement in math.

SECTION NINE

Interacting with Your Students

Each school day, the greatest portion of your time is spent interacting with students. You may be providing instruction, directing students as they use manipulatives to explore mathematical concepts, correcting inappropriate behavior, helping a student with his makeup work—the interactions throughout the day are seemingly as endless as they are varied.

Your interactions with your students will greatly influence their opinions of you and will affect their behavior in class. Students who view their teachers as being professional, confident, and caring are usually more involved in classroom activities and more focused on their work than students who perceive their teachers to be lacking these qualities.

Establishing positive interactions with your students must be one of your priorities. You must be willing to get to know your students, build trust with them, guide them to success in math, and help them to cope with the stresses that are an inevitable part of growing up. By interacting positively with your students, you will be teaching skills that will help them to achieve success in your math class, other classes, and in other areas of their lives.

Maintaining a Professional Role with Your Students

Ask students which teachers are the true professionals in their school, and they will quickly identify teachers they respect for their fairness, willingness to help, and ability to teach. All of these teachers project a strong, positive presence in their classrooms by the way they look, the way they speak, and the way they act. Possessing "teacher's presence" is a reflection of your professionalism.

YOUR APPEARANCE

The first thing people notice about someone is his or her appearance, which often leads to assumptions about the person. An individual who is dressed fashionably and is well groomed projects an image quite different from one who is poorly dressed and unkempt. Without question, students, parents and guardians, as well as staff members, will make assumptions about you based on your appearance. One of the most important ways to establish a professional teacher presence in the classroom is to look the part.

Though you should choose clothing in which you are comfortable, you should also avoid extremes that call attention to your style at the expense of professionalism. Select clothing that places you among the other professionals in your school. The clothing you wear should always be clean, wrinkle free, and appropriate for school. If your school has a dress code for teachers, be sure to adhere to it. Disregarding the dress code—for example, wearing flip-flops or jeans in violation of the code—sends a subtle message to your students that rule-breaking is acceptable.

In addition to dressing professionally, you should pay attention to personal hygiene. Make sure your hair is neat, your teeth are clean, and the smell of your breath and body are fresh. Never go to school smelling of cigarettes or alcohol, or overly strong perfume or cologne. Not only may some students find strong fragrances offensive, some may be allergic to them.

Presenting a professional image to others is the first step to being viewed as a professional. Teachers who ignore the importance of appearance risk undermining their professionalism in the eyes of students and of other adults in their school.

YOUR LANGUAGE AND TONE

A common complaint of teachers everywhere is that students do not always pay attention to what teachers say. This may be true when it comes to instruction, but your students probably hear much more than you think. What you say and how you say it can have a great impact on your students' impression of you, their feelings about math, and their opinions about school.

When speaking with your students, remember the following:

- Speak in a caring tone. Avoid harsh words and sarcasm.
- Express yourself clearly. Use vocabulary that is appropriate for the age of your students.
- Vary your intonations and facial expressions when talking. Avoid speaking in a monotone.
- Always use appropriate language, speak with correct grammar, and avoid slang.

- Use correct terms and math vocabulary when teaching and speaking with students about math.

- Pause periodically to give students a chance to process what you are saying.

- Be careful not to use words that some students may misinterpret. Subtle jokes that may amuse most students might hurt others. Avoid kidding about topics some students might consider to be sensitive.

- When you must correct the behavior of students, explain why the behavior was inappropriate. Avoid anger, criticism, and scolding.

The language and tone you use will enhance your professional image and foster communication between you and your students. Students respond more positively to teachers who speak to them in a professional and courteous manner.

YOUR BEHAVIOR

The way you behave with your students is as important as what you say to them. Because students look upon their teachers as role models, your actions will set the norm for what is acceptable behavior in your class.

Along with following the classroom rules you establish for your students, you should consistently demonstrate these behaviors:

- Encourage your students to be the best they can be.
- Listen to the concerns of your students.
- Try to help them resolve any problems they may have.
- Show patience with students, faculty members, support staff, and administrators.
- Express appreciation for the efforts of your students.
- Remain calm in difficult situations.
- Treat everyone with respect and consideration.
- Be organized.
- Be on time for class, assemblies, meetings, and special events.
- Be willing to laugh at yourself.
- Interact in a professional manner with students, parents and guardians, and the other members of your school community.

Your behavior in school should never slip into the category of "Do as I say, not as I do." Instead be sure your behavior is an example that students can use as a guide for conducting themselves properly in class.

AVOIDING POTENTIAL TROUBLE

As a teacher, you are in the public eye and may be an occasional topic of discussion among students, parents, guardians, and community members. You must be vigilant that what you do can not be misinterpreted. Gossip can be hurtful and, unfortunately, in our society harmless actions can quickly become major incidents. You must avoid situations that might be taken out of context and force you to defend what in fact was innocent behavior on your part.

It is wise to avoid the following and similar situations:

- Giving a student a ride home. For example, after a math club meeting, a student has missed the bus. Rather than offering a ride home, inform the office so that other arrangements can be made.
- Allowing students to visit you at your home.
- Visiting a student at his or her home.
- Tutoring a student at home or providing home instruction without a parent or guardian present.
- Helping a student before or after school alone with the classroom door closed. If you are helping a student, work at a desk that is visible through the doorway and leave the door open.
- Hugging or touching students, or displaying signs of affection.
- Addressing students with pet names.
- Posting questionable or compromising content, photos, or videos about yourself or others on the Internet. If you post information on the Internet, you can be sure students will find it. Note: If you post photos or videos of students at school events—for example, a math contest—make certain you are in compliance with your school's policies and secure all necessary permissions.

Always try to avoid situations in which you may be called upon to explain or justify your actions. Considering how others might perceive your actions can provide you with insight for making sound decisions when interacting with your students.

DISCRETION AND YOUR PERSONAL LIFE

Your students will be curious about you and will have numerous questions about your life outside of the classroom. Some of these questions will be simple and ordinary; some will surprise you; a few will shock you. You need to have a response for all of them.

While sharing some of your life is fine, for example, you enjoy skiing and have a pet cat named Noodles, it is not usually advisable to give students information such as your personal phone number, your personal e-mail address, or the address at which you live. Nor should you answer questions that ask your opinion of Mr. Sanchez, who your students believe is unfair and gives far too much homework every night. Anticipating the kinds of questions students may ask you will help you to decide in advance the kind and amount of information you are comfortable in sharing. As a general rule, any information you would feel uneasy about sharing with parents or guardians or administrators should not be shared with students.

Carefully consider responses to questions like the following before providing answers:

- Who do you live with?
- What is your family like?
- What do you do on weekends?
- Do you drink?
- Were you ever drunk?
- Do you smoke?
- Have you ever done drugs?
- What rules did you break as a kid?
- Were you ever in a fight?
- Were you ever suspended from school?
- Did you always do your homework?
- Did you ever lie to your parents?
- What's the worst thing you ever did?
- You had my brother last year. Isn't he stupid?
- My mom says I can't go to Monica's party. Do you think she should let me?

Sometimes students will press you for an answer you should not give. Do not allow them to pester. In such cases, simply say, "We're not here to learn about me; we're here to learn math," then move on.

Teaching is the professional part of your life. Your personal life is another part. Although they are intertwined, make every effort to leave your personal life at home during the school day.

How you dress, speak, and behave directly affects how your students perceive you. When you dress, speak, and behave in a professional manner, you convey to your students that they and your math class are important to you.

Getting to Know Your Students

As the school year progresses, you will learn more about your students, their personalities, learning styles, and academic strengths and weaknesses. You will also become aware of events and activities in their lives that can affect their performance and behavior in school. As you get to know your students, you will be able to better plan effective math lessons and manage your class for the greatest benefit of all.

You can start learning about your students before the school term begins. Checking their records and speaking with their former teachers can provide you with valuable information that may help you meet the needs of your students. The drawback to this is that past performance and behavior in another teacher's class is not always a good indicator of how a student will work and behave in your class. Avoid prejudging any student because of what he or she did in previous years. Every student deserves a fresh start to the new school year.

A helpful way to gather basic information about your students at the beginning of school is to ask them to complete a student inventory such as "Facts About You" in Section Six. This worksheet and others like it provide basic information about students that will help you to see them as individuals.

As you interact with your students and observe them as they interact with each other, you will discover that your class is filled with numerous personalities. The way your students meet the requirements of your class, work with their classmates, and conduct themselves in school will tell you much about them.

COMMUNICATION AND BEING A GOOD LISTENER

Because you must present a lesson in each period that includes instruction, directions, guidance, and a variety of activities, you will generally speak more to your students than listen to them. However, it is through communication and listening in particular that you will truly come to know the needs and concerns of your students.

Effective listening is a skill. It requires that you not only hear what another person is saying but also understand what he is saying. The actual meaning of a student's words may be straightforward—the student who explains that he did not complete his homework because he left his math textbook in school—or ambiguous—the student who justifies not doing his math homework because he does not have time due to his part-time job, when in fact his weak math skills make the work very difficult. Effective listening requires effort.

Following are suggestions for improving your listening skills as you interact with your students:

- Encourage your students to speak with you, sharing their ideas and opinions.
- Always take time to listen to your students. If you ignore them or imply that you do not have time for them, they will eventually stop sharing with you.
- Give students your full attention when they are speaking to you. Make eye contact.
- Focus on what is being said.
- Be aware of both verbal and physical cues. For example, a student who tells you that he does not care about his math grade but who is clearly

upset when he does poorly on a test may care more than he wishes to admit.

◎ Listen with empathy to your students. Put yourself in their position.

◎ Allow students to finish speaking before you speak. Refrain from formulating a response while the student is still speaking.

◎ Ask questions if you are not certain of what a student is saying. For example, comments such as "I'm not sure what you mean by that," or "Can you give me an example?" can be helpful in clarifying confusion.

◎ Give feedback with a nod or smile, and offer responses such as "I understand what you're saying," or "It seems that you're saying . . ." to assure students that you are listening.

◎ Respect what students say to you in confidence. Do not share what a student tells you with other students. (Note: If you are worried about the safety of a student or the safety of others, you must report your concerns to the proper administrator.)

◎ Try to speak with every student each day. Just a hello shows students that you know that they are in your class, and can be the opening of a conversation.

In addition to taking steps to listen effectively to your students, you should guard against actions that show your students that you are not willing to listen to them. Avoid the following:

◎ Doing another task—for example, grading papers when students are speaking to you at your desk

◎ Not paying attention

◎ Making judgments

◎ Being critical

◎ Rushing to conclusions

◎ Interjecting your own thoughts

◎ Giving a lecture

◎ Changing the topic by moving on to something else

◎ Minimizing the feelings of your students

◎ Offering a quick, simplistic solution to a complicated problem

Even as you work to improve your listening skills, you should work to help your students communicate with you (and others) effectively. It is probable that many of your students who are able to talk without seeming to stop for a breath may lack skills to communicate their ideas clearly. This can be particularly true when they are speaking about math.

To foster the communication skills of your students, hand out copies of the following "Tips for Positive Communication in Math Class."

Tips for Positive Communication in Math Class

Name _____ Date _____ Period _____

Communication is a two-part process: speaking and listening. Following are some tips for sharing your ideas and listening to the ideas of others.

1. Organize your thoughts before speaking. You must know what you want to say before you can say it.

2. Speak clearly. Get right to the point.

3. Look at the person to whom you are speaking. This will help her to listen and understand you better.

4. If there is a disagreement, be willing to compromise.

5. Give your full attention to others when they are speaking.

6. Do not interrupt others. Let them finish what they started out to say.

7. If you are not sure of what others mean, ask them to explain.

8. Be willing to accept constructive criticism during a discussion.

9. Think about how you will respond to someone before you start talking.

10. Be patient and polite.

11. Be positive. Do not:
    - Lose your temper
    - Yell or shout
    - Call people names
    - Be sarcastic
    - Belittle or insult others
    - Threaten others

12. Consider the following tips when talking about math:
    - Use proper terms and vocabulary.
    - Before asking your teacher for help with a problem, try to identify what part of the problem is confusing you. Try to be specific.
    - Do not be afraid to ask questions.

Even for students who possess relatively good communication skills, there will be times when they will have trouble starting or maintaining a dialogue. The following prompts can help you guide your students to express their thoughts:

- Can you tell me more about your problem?
- How have you tried to solve it?
- What do you want to do next?
- Based on what you know, what can you predict about . . . ?
- Can you find more information?
- Why did you decide to do . . . ?
- Have you solved a problem that is similar to this one?
- What might be the consequences of your action?
- How did you reach that conclusion?
- What other strategies might you try?
- How does this compare with . . . ?
- What do you think will happen if . . . ?
- So, are you saying that . . . ?
- Does this solution make sense?

The benefits of getting to know your students as quickly as possible are significant. When you understand your students, you are better able to adjust your classroom management to reduce potential behavior problems, create an effective learning environment, and plan and deliver math lessons that address the needs and capabilities of all your students.

Guiding Students to Become Successful Math Students

Underlying your interaction with your students is the primary goal of guiding them to be successful in their study of mathematics. By providing classroom experiences that encourage sound study skills and support students in their efforts to become responsible, active learners, you will give your students the opportunity to develop the behaviors that will assure their success in math, not just in your class but also in other classes.

To help your students recognize these behaviors, hand out copies of "How to Become a Successful Math Student," which follows. Discuss the tips with your students and suggest that they work to acquire these behaviors.

Few students are truly gifted in mathematics, but many students earn high grades. Most of these students have developed work habits that enable them to realize their full potential.

How to Become a Successful Math Student

Name _____ Date _____ Period _____

The following tips can help you become a better math student.

1. Report to class on time. Bring all of your materials and supplies, including your textbook, math notebook, agenda or assignment pad, and pencils.

2. Follow directions and class rules.

3. Start work promptly and participate in class activities.

4. Take notes.

5. Keep all of your math papers in your binder. Keep your materials organized.

6. Remain in class except for emergencies. For example, use the lavatory and make guidance appointments during your free time.

7. Work cooperatively with other students.

8. Be polite. Listen to what others have to say before speaking.

9. Ask questions if you do not understand a concept, procedure, or skill.

10. Use correct terms when discussing math.

11. Use various strategies to solve problems. Do not give up.

12. Use technology such as calculators and computers to solve problems or find information.

13. Complete all class work and homework on time.

14. Double-check your work.

15. Write down your assignments in your assignment pad or agenda.

16. Make up work you miss when you are absent.

17. Review your notes, classwork, and homework each night. This will help to keep math concepts and skills fresh in your mind.

18. Budget your time and plan ahead. If you know that you have a math project due on Friday, do not wait until Thursday to begin it.

19. Prepare for tests and quizzes. Complete reviews. Check your notes and classwork. Do some practice problems, especially those you find challenging.

20. Think of the big picture by relating math to other subjects and your life.

HELPING STUDENTS DEVELOP MATH STUDY SKILLS

All students are encouraged to study, but most are not taught how to study. This is especially true for mathematics. Yet, studying is the best way for students to achieve good grades in any subject. By presenting your students with effective steps for studying, you will be providing them with ways they can excel in your class, and in others.

To help your students develop math study skills, hand out and discuss "How to Improve Your Math Study Skills," which follows.

How to Improve Your Math Study Skills

Name _____ Date _____ Period _____

The following study tips can help you achieve success in math class.

1. Read your math textbook. Read headings and captions; study explanations and examples.

2. Pay close attention to the material your teacher presents in class.

3. Take notes in class. Here are some note-taking tips:
    - Write down concepts, main ideas, and formulas.
    - Write down important details and examples.
    - Pay attention to phrases that tell you that important information is to follow, such as "Remember this...," "The most important...," and "Don't forget to...."
    - Do not try to write down everything your teacher says. You will not be able to and you will probably miss important information.
    - After class, review your notes while they are fresh in your mind. Add any thoughts or ideas that may help you to remember important facts.

4. Practice by doing examples and creating problems of your own. You may also do practice problems with a study buddy, or complete practice problems on the Internet. (Check with your teacher for Web sites.)

5. Set aside a place and time for doing your homework. This should be a quiet place with sufficient lighting. It should be a place where you can concentrate.

6. When you are studying or working on homework, take a short break every twenty to thirty minutes. Periodic breaks will give your mind a rest, help you to recharge your energy, and allow you to return to your work refreshed.

7. When you make a mistake, try to figure out what you did wrong. If you cannot find your mistake, ask your teacher for help. Correcting mistakes leads to learning.

8. Make up any work you missed as soon as possible. If you are absent, try to get assignments from a friend so that you do not fall behind.

9. Study for tests by completing practice problems and reviewing your notes.

10. Remember that by working hard you can be successful in your math class.

HOW TO READ A MATH TEXTBOOK

Ask your average math student if he reads his math text, and his probable answer will be "Why?" Most students use their texts only to view example problems and to complete their homework. But a math text offers substantial information for mastering concepts and skills. Every chapter of the typical text has an introduction that puts the chapter topic into focus and highlights the material to come. Each section of the chapter follows with an objective and offers step-by-step explanations and examples that show students how to solve the problems that satisfy the lesson's objective. Chapters usually conclude with review pages or chapter tests that students can use for additional practice. The texts also include a variety of resources such as lists of formulas, units of standard measurement, symbols, technology tips, links to Web pages, an index, and a glossary where students can find the precise meaning of math terms.

Unfortunately, most students do not read their math texts. They may view examples and diagrams in their texts as they do their work, but they seldom read the accompanying explanations. They may not know how. Reading a math text is quite different from reading general interest articles, nonfiction books, novels, and texts in other subject areas.

To help your students read their math texts effectively, distribute copies of the following "Guidelines for Reading Your Math Text." Discuss the information and emphasize that reading their texts will help them to understand new material.

When students read their math texts, they are better prepared for class. They gain an appreciation of how notes, homework, and the activities of each day relate to the large concept of a unit of study and the application of math in their lives and other studies. Being able to read technical material, such as that in a math text, is a valuable skill that extends beyond the math classroom.

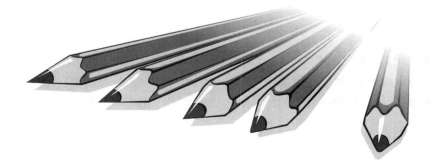

Guidelines for Reading Your Math Text

Name _____ Date _____ Period _____

Your textbook is one of your most important resources for your math class. The following tips will help you to read and understand your text.

1. Read the chapter or section summary of the material you are currently studying.

2. Look at the objective written at the beginning of each section. This will help focus your thoughts on the material of the section.

3. Skim the section you will study. Notice the main ideas. Skimming prepares your mind for a closer look at the material.

4. Read the section slowly and carefully. Pay attention to information that is highlighted. Make sure you understand new terms and their definitions. If necessary, check the glossary of your text for definitions. You may need to reread the section two, three, or more times to fully understand the ideas.

5. Take notes as you read to help you remember important concepts.

6. Carefully study examples. Be sure that you understand each of the steps of the example problems. If you are unsure about any of the steps, ask your teacher for help.

7. Carefully study any charts, diagrams, figures, tables, and graphs.

8. Work through practice exercises to test your understanding of the material.

9. Make certain you understand the specific meaning of any symbols that are used.

10. Practice using any calculator keystrokes that are introduced.

11. Visit Web sites that are suggested.

12. Ask yourself if you have satisfied the objective of the section. If you have not, reread the material to gain a better understanding. If you have, review the material to reinforce the skills and concepts in your mind.

HOW TO PREPARE FOR AND DO WELL ON MATH TESTS

Because they do not know the specific problems that a test contains, many students feel that it is impossible to study for math tests. Of course, this is not true. Tests assess mathematical knowledge, problem-solving skills, applications, the ability to use technology, and reasoning skills—all of which can be reviewed and sharpened before the test.

Your students need to realize that they can study for math tests. The following reproducible, "How to Prepare for Math Tests," shows them how. Distribute and discuss the information provided with your students.

How to Prepare for Math Tests

Name _____ Date _____ Period _____

The following tips can help you study for math tests.

1. Begin studying for the test a few days ahead of time.

2. Find out all you can about the test. If a study guide is provided, be sure to use it. If you are allowed to use a math reference sheet during the test that includes formulas or equivalencies, do not spend time memorizing this information. Instead focus on how to use the information.

3. If you are permitted to use a calculator, do example problems on your calculator.

4. Study with a friend. Explaining information to each other will help you to remember better. Do problems together and check each other's work.

5. Review problems from your homework and classwork.

6. Review the chapter in your textbook. Do at least one problem from each section. Take the chapter test in your text.

7. Anticipate new or tricky questions. Be ready for them.

8. Ask for help with problems you do not understand.

9. Think positively. View the test as a chance to show how much you know. Imagine doing well.

10. Have everything you will need: pencils, erasers, calculators, contacts or eyeglasses.

Preparing for a math test is the first step for your students doing well. The next step is to complete the test with confidence and skill. Hand out copies of the following "Math Test-Taking Tips" to give your students some suggestions on how they may take math tests effectively.

Math Test-Taking Tips

Name _____ Date _____ Period _____

The following tips can help you improve your score on math tests.

1. Review the test before you start working. Estimate how long each section should take you to complete. Budget your time. Do not spend too much time on any one problem.
2. Read and follow the directions carefully.
3. Read word problems carefully, and underline key words. Key words might indicate important data, relationships, or operations to use.
4. Show your work when you solve problems. Showing your work allows you to go back and check your math.
5. Work carefully. Avoid careless errors.
6. Do not become overconfident and rush through the material because you think it is easy.
7. Make sure that your answers to the problems make sense.
8. If you have an answer sheet, be sure to fill in the correct spaces. Make your marks dark.
9. Skip difficult problems. Come back to them after you have finished the rest of the test.
10. Here are some special tips for taking tests that have multiple-choice, true-false, and matching questions:
    - Read all of the answer choices for each question.
    - Reread a question if you do not understand what it is asking.
    - If you cannot decide between two or more answers, compare them. Eliminate the ones that seem less likely to be correct.
    - If you identify a statement as true, then it is true with no exceptions. If any part of a statement is false, the whole statement is false.
    - If you are working on a matching portion of a test, find out if a term can be used more than once. If it can be used only once, you can use the process of elimination to help you complete the section correctly.
11. If time remains, double-check your work.
12. Be sure that you have answered all of the questions completely.

How to Conduct Successful Math Conferences with Students

A math conference during class occurs when you meet with a student or students to talk about math. Such a conference does not have to be formal, and may last only a few minutes. It may take place at your desk, a student's desk, or a table at the back of the room.

The purpose of the conference is to help students with their understanding of math and their achievement in your class. Conferences may address various problems or issues, for example, mastery of a specific concept or skill, discussion of work habits, calculator use, a journal entry, or suggestions for selecting material for a math portfolio. During a conference you can reteach a skill, provide encouragement, and offer praise. You may also discover that several students are having trouble with the same kinds of problems and address those problems during a general review for the benefit of all. You may schedule math conferences periodically throughout the year, perhaps once each marking period, or you may hold conferences as frequently as you feel necessary.

The following tips can help ensure that the math conferences you have with your students will be successful:

- While you are meeting with individual students or small groups, be sure that the rest of the class is engaged with an assignment.

- Begin a math conference by asking the student how he feels he is progressing in math. If you are concerned that he is having trouble in a particular area, direct the conference to address your concerns.

- Address only one or two skills or problems during the conference. Trying to discuss several items can overwhelm students. Rather than leaving the conference with a positive outcome, they will leave the conference feeling frustrated or confused.

- Focus the conference to address the individual needs of each student or a small group of students.

- Listen to your students as they speak about their work in your math class. Their words and body language can help you to understand how to direct the conference for their greatest benefit.

- Be positive during the conference. Avoid negative comments. The conference should be a time of support, not of reproach or reprimand.

- Maintain notes that detail actions you will take to help the student. Be sure to follow through.

Math conferences allow you to interact with your students individually or in small groups. They present a wonderful opportunity for you to get to know your students, their strengths and weaknesses in math, and their concerns about their progress in your class.

The best math lessons and instruction will inevitably fall short if your students lack the necessary study skills. Helping them improve their skills and develop good habits for learning math is one of the most important tasks you can undertake during the year.

Preventing and Dealing with Disruptions

Because every classroom is filled with students with various personalities, each with his or her individual abilities and needs, keeping any classroom functioning smoothly requires understanding and effort. From resolving a minor dispute between two students over the use of a computer to explaining to a student why calling others names is unacceptable behavior, you must remain in control of your classroom throughout the day, every day.

The classrooms of teachers who are proactive in dealing with potential disruptions usually are more orderly than those of teachers who react to disruptions. Prevention of a disruption, clearly, is a better strategy than addressing the problem after it has occurred. You can prevent many disturbances in your class by getting to know your students. Understanding them and the stressors in their lives can help you to guide them through situations that might be frustrating and lead them to misbehave. A common example here is to avoid seating two students who share a mutual dislike next to each other. Despite your well-intentioned attempts to convince them to work together, their feelings might be too strong and may undermine their ability to learn and behave appropriately in class.

There is much you can do to prevent disruptions in your class. Consider the following strategies:

- Monitor your students. Circulate around the room to keep them on task. Try to be aware of what your students are doing (or not doing) even when you are not watching them directly, for instance when you are writing on the board and not facing them. Glancing back occasionally can aid your monitoring efforts. Consistent monitoring is one of your best strategies for minimizing disruptions and helping your students remain focused on their work.

- Act as a role model in class. Show respect, patience, and good manners.

- Plan interesting math lessons and diverse classroom activities. Do not allow downtime during class or at the end of the period in which students have little to do.

- Establish clear rules for your classroom.

- Enforce rules consistently and fairly. When students understand the rules of the class and see that you uphold the rules equally for everyone, they will respect you.

- Pay attention to all your students. Try to make every student feel that he is an important member of the class.

- Offer praise and encouragement to everyone in class.

Despite your best efforts to maintain an orderly classroom, disruptions will occur. When they do, intervene quickly to contain the problem. Remember that disruptions do not only affect the students directly involved but also the other students in the class. You must contain and manage disruptive events for the benefit of all.

When speaking with disruptive students, keep in mind the following:

- Focus on the behavior that needs to be addressed, not on a student's personality.

- Do not discuss the behavior in front of the class or other students. Speak privately with the student in the hallway; however, position yourself so that you remain in the doorway and can continue monitoring the class.

- Remain calm and objective.

- State precisely what behavior is unacceptable and explain the appropriate behavior.

- Use a firm but friendly voice.

- Do not overreact or lose your temper.

- Avoid putting the student on the defensive, as this will only block communication and make resolution harder.

- Do not allow yourself to become flustered, which will only result in making the problem more difficult to solve.

- Do not say things that you cannot carry out.

- Implement a plan or contract if the problem persists.

- Deal with disruption in accordance with your classroom and school rules.

When trouble arises in class, your students expect you to manage and rectify it. All students deserve a safe, orderly classroom in which learning is the priority. (See Section Thirteen, "Managing Inappropriate Behavior.")

Helping Students Cope with the Pressures of Being a Student

As you get to know your students, you will become aware of them as unique individuals. You will learn about factors and events that affect their behavior and schoolwork. You will come to know about stressors in their lives and may find yourself listening to their problems and offering advice.

Students may be stressed about many things, some of the most common being:

- Doing well in math (and other classes)
- Friends, fitting in with a group of other students, or resolving conflicts
- Relationships, especially with boyfriends and girlfriends
- Disorders such as severe allergies, or hearing, visual, or physical impairments
- Self-concept, for example feeling that they are too short or tall, overweight, or not as pretty or handsome as they would like to be
- Lack of money, which forces them to hold a job after school
- Feeling overwhelmed; not having enough time because they are involved in too many activities
- Parental pressure to excel in school
- Peer pressure
- Making career or course choices
- Illness or death of a relative or friend
- Family problems
- Disappointment, for example not being chosen for the cheering squad, football team, or lead in the school play
- Depression, which may be a result of several factors or conditions

Although many of these stressors may remain hidden in the background, their influence—usually negative—will often manifest in school. A good example is the average student who suffers extreme math anxiety because his parents demand that he earn straight A's in your class. Even though he does well, he believes he has never done well enough.

When your students come to you with problems, listen to them. (See "Communication and Being a Good Listener" earlier in this section.) Sometimes simply talking to an adult they trust helps students work through problems on their own. Though you may offer guidance from your own experiences, if appropriate, avoid giving advice that you are not qualified to give. Never minimize the problems of students, and do not hesitate to refer serious problems to the proper administrator or support staff.

SERIOUS PROBLEMS REQUIRING IMMEDIATE ACTION

While many of the stressors that affect your students are common to young people and can be resolved over a period of time, critical problems that may put students in danger require immediate action. Always follow your school's policies when dealing with problems that require urgent intervention.

Following are examples of serious problems that require your immediate attention:

- *The student under the influence of alcohol or illegal substances.* If you suspect that a student's behavior might be an indication of alcohol or drug use, follow your school's guidelines for action, which likely includes notifying your school nurse and an administrator. Do not accuse the student based on your suspicions. Write down notes detailing the behavior you observed. Be as accurate as possible. If necessary, remove the student from class.

- *The student suffering a medical emergency.* If a student experiences a medical emergency, for example, passes out in class, suffers a seizure, or has a severe allergic reaction, you must act promptly. Call or send a student for the nurse and make the ill student as comfortable as possible. This may include moving desks aside so that the student may recline on the floor. Remain calm and instruct other students to step back and provide room. Do not administer medications or first aid unless you are trained and authorized to do so.

- *The pregnant student.* If a student confides to you that she may be pregnant and asks you what she should do, be extremely cautious in offering advice. First, ask if she has told her parents or guardians of her concerns. If she has, suggest that she seek their advice. If, however, she has not informed them, or is afraid to inform them, suggest that she speak with a guidance counselor who can refer her to the appropriate professionals. You might want to accompany her to the guidance office in an offer of support. If possible, inform the guidance counselor in advance that you are coming.

- *The abused student.* Students may be abused by family members, family friends, boyfriends, girlfriends, other students, or neighbors. If a student mentions that she is being abused, or if you suspect abuse is occurring—you notice bruises and her explanation is inadequate—speak to her after class. If she admits to being abused, or despite her denials you still suspect abuse, report the incident to the proper administrator in your school. Then follow your school's or state's protocols for reporting suspected abuse.

- *The suicidal student.* If a student speaks about suicide, or even in passing talks of harming or killing himself, or if you hear about a student who has talked about killing himself, you must not take such comments lightly. Let the student know that you are concerned about him and make immediate arrangements for him to meet with a guidance counselor or administrator. Wait with him until a counselor or administrator arrives. Do not leave him alone. You must follow through, even if the student now claims he was only kidding. The counselor or administrator must determine the degree of danger, if any.

- *The student who may harm others.* If a student threatens to harm or kill someone, or talks about bringing weapons to school, you must immediately report this to an administrator and follow your school's policy regarding such threats.

Whenever you must manage a crisis in school, try to remain as calm as possible. Take a deep breath to steady yourself. Not only will a calm demeanor on your part have a reassuring effect on students, it will enable you to think clearly and exercise better control over the situation.

Quick Review for Interacting with Your Students

The relationships you build through the daily interactions with your students can have a significant impact on their academic success and personal growth. Students who have good rapport with their teachers generally go on to enjoy satisfying and productive school years.

The following guidelines summarize how, through positive interactions, you can build relationships marked by mutual trust and respect with your students:

- Be professional at all times. Your appearance, the way you talk, and how you act affects how others perceive you. This in turn influences your relationship with them.

- Take time to get to know your students—their personalities, interests, strengths, and weaknesses—and create a positive classroom environment that best supports their learning needs.

- Learn to be an effective communicator and listener. Be one of those special adults in your students' lives who truly hears what they have to say.

- Guide your students in developing productive work habits and good study skills for learning math.

- Periodically conduct math conferences with your students. Conferences provide an opportunity to reinforce skills and concepts, offer suggestions that foster learning, and give encouragement and praise.

- Help students cope with the stressors in their lives.

- Work to prevent disruptions in class. When disruptions do occur, address them promptly and effectively.

- Act immediately when serious problems occur. If necessary, enlist the assistance of administrators and support personnel.

For many teachers, interacting with their students is one of the most enjoyable parts of their job. As teachers learn about their students, they are better able to help students achieve academic and personal success. It is always a great pleasure to see your students acquire new skills that will help them attain their goals.

Designing Effective Math Lessons and Activities

Although many factors affect the learning of the students in your class—including their abilities, interests, and behavior—one of the most important is your planning effective lessons and activities. Planning should be a priority for each school day.

When you plan, you become the designer of your course and determine the direction and destination of learning. Effective planning will help enable you to spark the interest of your students, tailor teaching techniques to satisfy different learning styles, and ensure the success of your teaching. Moreover, it will give you the opportunity to develop lessons and activities that precisely address state standards and district goals, as well as the objectives of your curriculum. Effective planning permits you to present individual concepts and skills sequentially and strengthens your course of study.

Whether you are creating a unit plan or a daily lesson, effective planning requires time, thought, and effort. But the rewards of sound planning are significant: a successful year for you and your students.

Making Time for Planning

Because planning is an ongoing responsibility, you must set aside time for it. Though some teachers prefer to plan in school before students arrive or after they have gone, others plan during preparation periods, and still others like to write their plans at home. Whichever way you choose is fine as long as you set aside a consistent time to plan. When you build planning for lessons and activities into your schedule, it becomes a part of your day. Avoid slipping into the practice of planning at irregular intervals, which will inevitably result in your falling behind and then rushing to catch up, only to fall behind again. Such an approach will lead to stress and plans that are likely to be mediocre at best.

The following tips will help you to make time for planning:

- View planning as one of your most important responsibilities as a teacher.
- Decide on a place and time for planning. Whether at school or at home, make planning effective lessons and activities for your students a part of your routine. This will help you to plan consistently.
- Always have handy all the materials you need for planning. This includes textbooks, teacher's manuals, curriculums, state standards, district goals, and supporting resources. Having to find a resource book for the review worksheet you need wastes time and adds frustration to your day.
- Have necessary technology available. For example, you may need a computer to obtain information from math Web sites, write your plans, and generate worksheets or tests. If the lesson you are planning requires students to use a calculator, you may find it helpful to review keystrokes, settings, and menus.
- Try to write your plans in one sitting. When you try to write them in bits—starting them in the morning before students arrive, doing a little more during your preparation period, and finishing them at home—you risk losing continuity. You will find yourself constantly reviewing what you did in an effort to move forward with the lesson. This will waste valuable time and may result in a weak lesson that reflects poorly on your teaching and leaves your students struggling with the material.

Effective planning helps you to create a positive learning experience for your students. It results in successful lessons and activities and is a critical factor in student achievement.

Planning a Variety of Math Lessons and Activities

For years, traditional math instruction consisted of the teacher standing at the front of the room, writing problems on the board and explaining the skills of the day's lessons, and then assigning practice problems for students to complete. This was usually followed by a homework assignment that provided students with more problems similar to those done in class. Today's math instruction is much more than the simple skills and drills of the past.

To plan for the success of their students, math teachers now must:

- Address the needs of various learning styles
- Provide problems that engage students' critical thinking skills

- Allow students to represent their work in a variety of ways
- Present complex problems that can be solved via different strategies and that, depending on the justification of answers, may have different solutions
- Present problems in real-life contexts
- Link math to other disciplines, students' lives, and global issues
- Use manipulatives for modeling mathematical concepts and relationships
- Incorporate technology into lessons and activities
- Integrate writing in the math curriculum
- Encourage students to verbalize their thoughts and evaluate their own work
- Utilize cooperative learning
- Allow time for reflection

When you plan a variety of math lessons and activities, you expand the learning experiences of your students. Variety will stimulate your students' interest and enthusiasm.

The Foundation of Successful Math Lessons and Activities

Successful math lessons and activities are built on objectives. These objectives are found in your curriculum and are usually aligned with state standards and district goals, as well as the Principles and Standards of the National Council of Teachers of Mathematics (NCTM). In order to design effective lessons and activities, you must be cognizant of the objectives that drive your curriculum.

STATE MATH STANDARDS

States have standards that identify the material students should master by the end of a particular grade or course. These standards are appropriate for each grade level and move sequentially from one grade to another. You should use state standards as the starting point for planning your lessons and activities.

To include state standards for math in your planning, do the following:

- Research your state's department of education Web site for math standards (or go to www.educationworld.com/standards/state for a list of standards by state). Print a copy for your math level or course. Keep this with your planning materials and refer to it as you plan throughout the year.
- As you plan your lesson or activity, be sure to address your state's standards.
- Keep in mind that some state standards are very broad. You may devote several days to teaching the skills and content of one standard. Sometimes you may be able to address components of several standards in one lesson.

The objectives that appear in state standards often appear on standardized tests, which are used to assess student proficiencies. Always refer to your state's standards for math when you write unit and daily lesson plans.

DISTRICT AND SCHOOL MATH GOALS AND OBJECTIVES

Just as states have standards for learning, your district and school have goals and objectives. Although these goals and objectives are often based on the standards, they may extend beyond those of your state. Being a member of your district and school's teaching staff, you are expected to support local goals and objectives in your instruction.

You can familiarize yourself with the goals and objectives of your district and school by checking your district's Web site or consulting with your principal or supervisor. You should plan lessons and activities that incorporate these goals and objectives whenever possible.

CURRICULUM

Along with outlining the content that should be taught, the typical curriculum provides time lines for instruction, objectives, suggested activities, and assessment tools. The curriculum is your guide for your math course, and you should use it as the basis for your lesson plans.

Here are some suggestions for utilizing your curriculum in the design of lessons and activities:

- Familiarize yourself with your curriculum before the school term begins.
- Break the curriculum down into units you will be teaching.
- Note any pacing guidelines. This will help you plan individual lessons and keep you moving forward at a realistic pace.
- Pay close attention to topic development. Develop material sequentially and logically.
- Prioritize lessons and activities so that students learn important material in the time allotted for a unit. You can always add supplementary or enrichment material if you have extra time.
- Collaborate with colleagues who are teaching the same course. Although the needs of classes differ, try to maintain approximately the same pace and cover the same core material. This helps to ensure that all students are exposed to the same concepts and skills.

Your curriculum provides a framework on which you can build a successful math program. It is a vital guide that can help you plan lessons and activities that satisfy the goals and objectives of your course.

THE PRINCIPLES, STANDARDS, AND FOCAL POINTS OF THE NCTM

The NCTM (www.nctm.org) is a professional organization for math teachers. Through its Principles, Standards, and Focal Points, the NCTM details the knowledge, understanding, and skills in mathematics that students in grades Pre-K through 12 should master. (See *Principles and Standards for School Mathematics*, NCTM, 2000, and *Curriculum Focal Points for Prekindergarten through Grade 8 Mathematics*, NCTM, 2007.)

Whereas the Principles describe specific elements of effective mathematics education, the Standards provide expectations of what students should know. The Standards are divided into two groups: content standards and process standards. The content standards, which often overlap and are interconnected, state what students should learn, and include Numbers and Operations; Algebra; Geometry; Measurement; and Data Analysis and Probability. The process standards, which describe methods for utilizing the material in the content standards, include Problem Solving; Reasoning and Proof; Communication; Connections; and Representations.

The Focal Points are specific clusters of mathematical concepts, skills, and procedures that should be addressed at each grade level through grade 8. Referring to the focal points during planning can help you design lessons, activities, and assessments that concentrate on the most important mathematical topics for your grade.

The NCTM also offers books, periodicals, and online support materials that include lesson plans, problems, and activities. The NCTM can be a valuable resource.

OBJECTIVES

The objective, or objectives, of every lesson or activity is the purpose of the lesson. Quite simply, objectives are what you want students to be able to do at the end of the lesson.

Objectives should be clear, specific, and measurable. They should relate to state standards, be included in your curriculum, and support district and school goals. Consider this example of an objective: Students will be able to use ratios of corresponding sides or scale factors to find missing lengths of similar figures. This objective is specific and clearly tells students what they are expected to do. It is also measurable. To satisfy the objective, students must successfully find missing lengths of similar figures. Once you have decided on an objective, you must develop the lesson or activity in a manner that enables students to master the skills needed to satisfy the objective.

You should always tell your students the objectives for each math class, and write them on the board in the same place every day. Knowing what they are expected to do helps to keep students focused on their work throughout the class.

Objectives help you to stay focused as well. Sound objectives are a means of ensuring that your students are exposed to all of the concepts and skills necessary to succeed in your math class.

RESOURCES FOR PLANNING MATH LESSONS AND ACTIVITIES

Along with your state standards, district goals, curriculum, and the Principles, Standards, and Focal Points of the NCTM, the following resources are useful for planning:

- *Your math text and teacher's manual.* Most math texts contain an enormous amount of information that can be the basis for lessons and activities. Your text may contain supplementary charts and tables, URLs of useful Web sites, calculator activities, extensions, ways for determining readiness, assessments, and lists of additional resources.

- *Math resource and reference books.*

- *The Internet.* On math Web sites you can find ideas, data, lesson plans, worksheets, and activities. Some particularly useful Web sites for planning include:

  - http://illuminations.nctm.org, which is part of the Web site of the NCTM.

  - www.educationworld.com, which offers a wide range of math lesson plans, math facts, and puzzles.

  - www.mathforum.org/library/resource_types/lesson_plans/branch .html, which offers information about math and lesson plans.

  - www.mathpropress.com/glossary/glossary.html, which contains definitions of math terms used in middle and high school math courses.

  - www.convert-me.com, which allows users to convert units of measurement.

  - You may also search for a specific topic that may result in Web sites that you can use in planning. For example, a search using the term "subtracting integers" will lead to many helpful Web sites. (For additional Web sites see "Resources on the Internet" in Section Three.)

- *Data from newspapers and magazines.* Articles in print form and online often contain charts and tables which can be used in developing math lessons and activities.

Every successful math lesson and activity begins with planning. When you include clear objectives based on state standards and district goals, your instruction will provide your students with the essential concepts and skills they will need for the satisfactory completion of your course.

Components of Effective Math Plans and Activities

Along with being founded on clear objectives and standards, effective math plans and activities share several components. They should address the needs of diverse learners, build on prior knowledge, include a variety of purposeful activities, promote critical thinking, incorporate technology (if possible), and provide a means for evaluation. Effective plans and activities move students forward in the topic of study.

ADDRESSING THE NEEDS OF DIVERSE LEARNERS

Every student is an individual with different interests, abilities, and learning styles. One of your challenges as a teacher is to create lessons and activities that meet the needs of all your students. An overriding goal throughout the year should be to help every student achieve success in learning math.

To address the needs of the diverse learners in your class, consider the following:

- Pre-assess your students. Review their records, standardized test scores, and previous math grades. Understanding their abilities in math can help you design lessons and activities that will be appropriate for their level. Consider giving your students a pretest for each unit.

- Differentiate your instruction, which essentially means using a variety of techniques and strategies to address the individual needs of your students. You can differentiate your instruction by considering each student's readiness, interests, and learning styles during planning. This kind of instruction allows you to plan lessons and activities that will engage your students with math. You can differentiate in three ways:
  - *Content:* Although topics, concepts, and skills should be the same for all students, the degree of complexity may be adjusted according to individual needs.
  - *Process:* Your plans for instruction—whole class, group, and individual—should take into account the abilities and learning styles of your students.
  - *Products:* Assessments should allow students options of how to demonstrate their understanding of a topic. Further, assessments should be designed to appeal to various learning styles.

○ Be familiar with your students' learning styles. Over the years, a vast amount of information has been written about the ways students learn. Although many different authors provide many different labels and breakdowns, resulting in much overlap, basic learning styles can be divided into three main types: auditory, visual, and kinesthetic.

- The auditory learner acquires information best by hearing. This student will do well with traditional teaching methods such as lectures, listening to explanations of the solutions to sample problems, and discussions of concepts, properties, and procedures.

- The visual learner acquires information best by seeing. These learners can be divided into two subgroups—verbal and nonverbal. While verbal visual learners tend to prefer information that is presented in written language, for example, explanations in texts, written lists, step-by-step procedures, and examples, nonverbal visual learners prefer to receive information in the form of charts, figures, diagrams, representations, and tables. Visual aids, such as number lines, charts showing equivalencies, and pictures of cylinders and cones, are easily understood by nonverbal visual learners.

- Kinesthetic learners prefer to learn via hands-on activities that include movement and touch. These learners like to tinker to figure things out. They benefit from using manipulatives, such as Cuisenaire® Rods and Geoboards, examining or creating models of three-dimensional figures, constructing geometric shapes, playing math games, and working on a computer or with a calculator. (See "Common Math Manipulatives" in Section Three.)

- Keep in mind that although a particular learning style may be dominant for a student, that student can also learn through other means. For instance, a student may be considered an auditory learner who learns best through hearing, but he may also benefit from using manipulatives, especially if students are talking about their discoveries and conjectures as they explore concepts with manipulatives. It is also important to realize that all students benefit from learning about a subject through different types of activities.

- Here is an example of how to differentiate instruction in a lesson about adding two integers. Explain the processes and provide numerical examples for your auditory learners. Write notes on the board or overhead, including the procedures and examples, for your verbal visual learners. Model adding integers on a number line or use virtual manipulatives to demonstrate the process for your nonverbal visual learners.

Finally, provide two-color counters for your kinesthetic learners to use to find the sum of two integers.

- For more information about learning styles, visit Learning-Styles-Online (www.learningstyles-online.com), or search the Internet using the term "learning styles, math."

◎ Provide alternative forms of assessment. Not every student is a good test taker. Rather than using a conventional test as an end of the unit assessment, you might instead assign a project that allows you to evaluate students' knowledge. You might also combine test results with math projects or portfolios to broaden the scope of your assessment. (See Section Twelve, "Evaluating the Progress of Your Students.")

◎ Include a variety of activities in class each day to stimulate the interest of your students and allow them to learn and explore math concepts and principles.

◎ Allow students to work in groups as well as individually.

◎ Know and implement the IEPs and 504 plans of your students.

◎ Provide enrichment for students who have mastered the content of a unit and are ready for an extension of the topic.

Your students are individuals. They have separate needs, abilities, and learning styles. When you design lesson plans and activities to meet the needs of the diverse learners in your class, you are ensuring that all students have an equal chance to excel, learn in their preferred style, and experience success that is measured with various assessment tools.

BUILDING ON PRIOR KNOWLEDGE

All of your students possess prior knowledge that they can apply to new topics. Understanding the prior knowledge of your students enables you to design effective math lessons and activities that address their interests and needs.

The prior knowledge of your students affects what and how you teach. If your students have a solid understanding of a concept—for example, if a seventh-grade class already understands (from sixth grade) equivalent fractions and decimals and can easily convert from one to another—you may find it unnecessary to spend as much time on these topics as you would for a class whose understanding is weak. For a class with a weak understanding of equivalent fractions and decimals, you should provide lessons that reteach these skills before moving on. Also, if you find that individuals or only a few students in a class are well advanced or severely lacking in skills, you should provide appropriate enrichment or remedial materials. Before you begin any unit of study, you should determine what your students know about the topics that will be presented.

To determine the prior knowledge of your students, you may do the following:

- Administer a pretest of the skills and concepts contained in the unit. Evaluate the test and find areas of strengths and weaknesses. Design your plans and activities to address the objectives your students need to master.

- Survey your students by asking them to list what they know about a specific topic. For a unit about multiplication and division of fractions, you might survey them about multiplying simple fractions, multiplying fractions and whole numbers, multiplying mixed numbers, dividing simple fractions, dividing fractions and whole numbers, and dividing mixed numbers. Use the results of the survey to guide your planning.

- Conference with groups of students and discuss what they know about a topic. Write notes about their understanding. Evaluate your notes and base your lessons and activities on your findings.

Your students' prior knowledge on topics provides you with guidelines for designing effective lessons and activities. Knowing what your students know about a topic, and what they need to know, allows you to focus the material in the most efficient manner.

INCLUDING MATERIAL FOR CRITICAL THINKING IN YOUR MATH PLANS

An important goal of math instruction is to help students develop critical-thinking skills that will facilitate life-long problem solving. Before they can think critically, however, students must have fundamental knowledge on a topic. Bloom's Taxonomy is an effective model to help you develop and ask questions that promote critical thinking in your class.

Bloom organizes thinking into six categories, which can be applied to mathematics:

- Knowledge—students recall information, formulas, facts, and procedures
- Comprehension—students understand the material and are able to explain it
- Application—students use a concept in a new manner
- Analysis—students separate a concept into smaller parts
- Synthesis—students combine concepts in a new or different way
- Evaluation—students make judgments

The goal of critical thinking is to help students make accurate evaluations. You can foster critical thinking by asking your students to recall, explain, apply, analyze, and synthesize information. Searching the Internet with the term "Bloom's Taxonomy, math instruction" will result in many helpful Web sites.

You can incorporate critical-thinking skills into your lessons and activities by doing the following:

- Include problems that may be solved in different ways and may have different solutions. In these kinds of problems, students must justify their results as well as consider the results and strategies of others.
- Provide real-life problems that tie into other subjects.
- Give students time to reflect on what they have learned.
- Ask open-ended questions in which students must explain their answers, such as:
  - How would you solve this problem?
  - What other strategies might you use to solve this problem?
  - What is the best way to solve this problem?
  - How might the results change if . . . ?
  - What can you infer from . . . ?
  - What is your estimate of . . . ?
  - What can you predict if . . . ?
  - What if . . . ?
- Have students evaluate their own work and that of their peers and provide suggestions for improvement.

Critical thinking is a vital aspect of effective math lessons and activities. By promoting higher-order thinking, you will help your students develop skills that will serve them well in all areas of their lives.

INCORPORATING TECHNOLOGY IN MATH PLANS

Your students are growing up in a technology-based world. They are certainly familiar with technology that is used for communication, entertainment, and music, but they also need to be comfortable and competent with the applications of technology necessary for learning and work.

Technology is undisputedly a necessary component of any math class. Technology—particularly calculators, computers, and Internet access—enables students to compute efficiently, explore mathematical ideas via visual representations, and organize and analyze data. Technology frees students from paper-and-pencil computation and allows them to channel their efforts on reasoning, problem solving, and decision making, helping them to develop higher-level skills.

Use the following guidelines when incorporating technology in your lessons:

- Make sure that your use of technology relates directly to the objective of the lesson. Technology should be a part of the lesson and enhance learning. It should never simply be an add-on. Most texts suggest calculator activities that you may use with your students.

- Utilize computers and calculators to explore mathematical concepts through simulations and dynamic software.

- Make certain that you are familiar with any equipment, software, simulations, or Web sites you will use with your lesson. Fumbling around trying to figure out keystrokes during the lesson will only frustrate you and make it difficult to achieve your lesson's objective. It will also confuse students and cause them to lose interest.

- Set up routines and procedures for using technology in your class. (See "Distributing and Collecting Materials" and "Procedures for Student Computer Use" in Section Seven.)

- Make sure that you have enough equipment available. If not, arrange for students to share.

- Always make sure that equipment is working properly. If you are using a Web site, check that it is operational. If you are using calculators, have enough fresh batteries on hand.

- Carefully plan any activity involving technology. Make sure that the level of difficulty is appropriate for your students, and that students will have enough time to complete the activity.

- Always have a backup plan. If a Web site is down, or software experiences a glitch, you should have an option available for completing your lesson.

By integrating technology into your lessons and activities, you are providing the tools that allow students to focus on solving problems, exploring principles, and sharpening their reasoning skills. Through the use of technology, you will be enriching your students' understanding of mathematics.

PROVIDING A MEANS FOR ASSESSMENT

One of your ongoing responsibilities as an educator is to assess your students' progress. Assessment should be reserved not only for tests, quizzes, and major activities such as math projects, but should be a part of every lesson. The information you gain from assessments enables you to measure and evaluate the learning of your students, and helps you to plan effective subsequent lessons.

You can assess daily lessons in a variety of ways, including:

- Checking classwork and homework.
- Monitoring the daily work of your students.
- Conferencing with students or groups of students.
- Using math do-nows or math starters at the beginning of class to check learning of the previous day's lesson.
- Asking questions throughout the lesson to check for students' understanding.
- Using an entrance card. Similar to a do-now, students write a response to a question based on the previous day's work or homework.
- Providing short quizzes on the previous day's work or homework.
- Using an exit card. After the day's lesson is complete, but before students leave class, ask them to write a statement about what they learned or answer a question.
- Using 3–2–1 cards. At the end of the lesson, instruct your students to write three things they learned, two things that are unclear to them, and one thing they have a question about.

You should use the assessment tools that are most appropriate for your class and for evaluating specific lessons. Including a means of assessment with every lesson enables you to know what your students have learned and what they still need to learn about specific topics, concepts, and skills. (See Section Twelve, "Evaluating the Progress of Your Students.")

All effective math lessons and activities have much in common. They address the individual needs and learning styles of students, foster critical thinking, utilize technology as often as possible, and always provide a means for assessment.

Types of Math Plans

Most math teachers are responsible for creating two types of plans: unit plans and daily lesson plans. Both are based on your curriculum.

Before you can write any plans, however, you must know exactly what your students must learn by the end of the school year. Your curriculum guide will be your greatest resource. Review the philosophy, goals, objectives, strategies, materials, time frames, and assessment tools that are suggested. In addition, check your state standards and district and school goals for your grade level or the subject you are teaching to make certain that you cover essential skills and concepts. Your textbook and supplementary resource materials will provide still more information you can use in planning and creating a comprehensive program that encompasses the required material at your students' grade level.

To help you plan your year, consider completing the "Basic Course of Study Planning Guide" that follows. Keep the guide in your binder for easy reference. For each month, simply list the major topics you intend to teach and the approximate length of time you expect you will need. By the end of the school year, you should be able to cover all of the major topics you must teach and possibly some additional ones you might use for enrichment or reinforcement. Following is an example showing one month of the guide.

September: Number of School Days <u>18</u>

Introductory Activities for the School Year (2 days)

Chapter 1, pages 2–44

Undefined Terms (2 days), Segments and Rays (3 days), Angles (2 days), Angle Pairs (3 days), Perpendicular Lines (2 days), Culminating Activity and Test Review (3 days), Chapter Test (1 day)

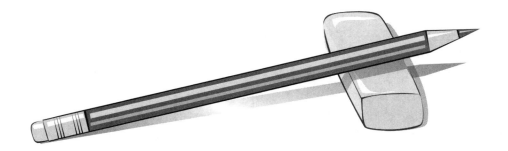

Basic Course of Study Planning Guide

Course Title: _____

August: Number of School Days _____

September: Number of School Days _____

October: Number of School Days _____

November: Number of School Days _____

December: Number of School Days _____

January: Number of School Days _____

Basic Course of Study Planning Guide (continued)

February: Number of School Days _____

March: Number of School Days _____

April: Number of School Days _____

May: Number of School Days _____

June: Number of School Days _____

July: Number of School Days _____

Completing the guide will help you to keep moving forward at a steady pace. If, for example, you find that by November you have covered more topics than you had anticipated, you may consider giving students enrichment or more challenging material. If you find that you are not as far along as you expected, however, you must either revise your upcoming plans or move ahead at a faster pace. Look upon the guide as a reference, and adjust your plans as necessary.

Think of your curriculum guide as a roadmap that shows you the general course you must follow from the beginning of the school year to its end. Because it summarizes what you will be teaching, it is an important tool for planning.

UNIT PLANS

Unit plans, which address major topics and related concepts and skills, are logical and sequential divisions of your curriculum. They provide the guidelines for writing daily lesson plans.

The material in a unit should share a common theme, the length of time needed for instruction should be realistic, and the activities should be student-centered rather than teacher-directed. For example, instead of telling students what they should do, activities should give them the opportunity to explore and discover mathematical concepts, methods, and procedures. The material in a unit should also take into account the prior knowledge, learning styles, and general abilities of your students.

The following steps can help you plan effective units:

- Identify the big picture by summarizing the main ideas of the unit. This will help you focus on development and presentation of the material.

- Formulate questions that address major concepts, and which require students to think critically. Such questions will guide you in the development of material that fosters higher-level skills and focuses students on the central themes of the unit.

- Formulate your objectives. What do you want your students to be able to do upon conclusion of the unit? Use state standards, district and school goals, your curriculum, text, and other resources to determine desired student outcomes. The outcomes must be measureable and can be used to determine student progress towards your program goals.

- Decide on content. What information will your students need to know to meet the objectives of the unit? This might include properties, vocabulary, procedures, operations, theorems or rules, and problem-solving strategies.

- Decide which activities and instructional strategies you will use to help your students meet the objectives of the unit. You should include a variety of activities that will interest students, address special needs and different learning styles, and provide opportunities for problem solving and higher-level thinking skills. Some activities you might consider include:

- Lectures
- Demonstrations
- Do-nows
- Making models or modeling concepts
- Exploring with manipulatives
- Using dynamic, interactive math software
- Textbook assignments
- Supplementary activities in resource books
- Calculator explorations
- Real-life investigations
- Exploring math Web sites
- Projects
- Presentations
- Cooperative group work
- Writing
- Journal entries
- Conferencing
- Responding to open-ended questions
- Enrichment or challenges

○ Decide which resources and materials you will use to present the content of the unit. These might include your textbook, supplementary reference and resource books, manipulatives, calculators, math software, computers, and math Web sites.

○ Include assessments. How will you assess the learning of your students for this unit? Assessments may take several forms, including checking homework and classwork, classroom observations, quizzes, tests, projects, or presentations. (See Section Twelve, "Evaluating the Progress of Your Students.")

Your school may provide its teachers with a format for writing unit plans. If you do not have a specific format, use the following "Unit Plan Format," or use it as a guide to design one that suits your particular planning needs. Consider creating an electronic template and store completed pages on your computer.

Unit Plan Format

Teacher: _____   Course: _____

Unit Title: _____   Dates: _____ to _____

I. Summary

II. State Math Standards

III. Questions

IV. Objectives

Unit Plan Format (continued)

V. Content

VI. Activities

VII. Resources and Materials

VIII. Assessments

The following sample unit plan on rational numbers for a seventh-grade class illustrates the major parts of effective unit plans. Note that the math standards will vary according to your state.

Sample Unit Plan

Teacher: *Mrs. Sanchez*   Course: *7ᵗʰ Grade Math*

Unit Title: *Rational Numbers*   Dates: *2/1/10 to 3/5/10*

I. Summary

Positive and negative numbers help us to understand and describe real-world situations. The operations of addition, subtraction, multiplication, and division are defined and modeled in several ways. The commutative property, distributive property, and order of operations are applied to real-life problems with integers and rational numbers.

II. State Math Standards

- *Compare and order numbers of all types.*
- *Use and explain procedures (pencil and paper, mental math, and calculator) for performing calculations with integers and rational numbers.*
- *Understand and apply the standard algebraic order of operations.*
- *Understand and apply the properties of operations, numbers, and equations.*
- *Recognize, describe, extend, and create patterns involving whole numbers, rational numbers, and integers.*

III. Questions

- *How do negative and positive numbers help us understand a situation?*
- *What methods can we use to add, subtract, multiply, or divide positive and negative numbers?*
- *What model(s) for positive and negative numbers show relationships in real-world situations?*
- *What properties apply to operations with rational numbers and how are they used to solve problems?*

IV. Objectives: Students will be able to:

- *Represent positive and negative numbers, opposites and absolute value.*
- *Compare and order rational numbers and locate them on a number line.*
- *Find the opposite and absolute value of a number and explain the relationship.*
- *Add, subtract, multiply, and divide positive and negative numbers using models, patterns, and rules.*
- *Write equations to show relationships.*
- *Use inverse operations to solve one-step equations.*
- *Apply the commutative property for addition and multiplication, and use a counterexample to explain why it does not apply to subtraction and division.*
- *Use the order of operations and distributive property to simplify expressions.*

V. Content

- *Integers (definition, representations, opposites, operations, and properties)*
- *Rational numbers (definition, representations, opposites, operations, and properties)*
- *Number line (location of numbers, ordering, comparing)*
- *Absolute value (distance of a number from zero, notation)*
- *Operations with integers and rational numbers (addition, subtraction, multiplication, division)*
- *Inverse operations (addition-subtraction, multiplication-division)*
- *Equations (definition, writing, and solving one-step equations)*
- *Properties (commutative, distributive)*
- *Order of operations*
- *Expressions (definition, simplifying)*

VI. Activities

- *Number line activity*
- *Using counters*
- *Number races*
- *Integer bingo*
- *Lot division*
- *Integer operation review*

- *Modeling solutions to simple equations*
- *Stock market project*
- *Note taking and maintaining a notebook*
- *Questions and short answers*
- *Do-nows*
- *Journal entries*

VII. Resources and Materials

- *Number lines*
- *Rulers*
- *Two-color counters (for student use and overhead projector)*
- *Transparencies, markers, and overhead projector*

- *Bingo sheets*
- *Spinners for integer operation review*
- *Newspapers*
- *Computers with Internet access*
- *Interactive whiteboard*
- *Calculators*

VIII. Assessments

- *Journal entries and do-nows*
- *Peer conference after integer bingo*
- *Student-teacher conference after lot division*
- *Quizzes*

- *Various homework and group work assignments*
- *Unit test*
- *Stock market project*

After creating unit plans, keep them on file. If you are teaching the same course the following year, you will be able to easily update your original plans.

Unit plans enable you to divide your curriculum into manageable segments. This in turn allows you to concentrate on specific topics, concepts, and skills within a given time frame. Unit plans are essential parts of the big picture of your teaching year.

DAILY LESSON PLANS

Daily lesson plans are developed from unit plans. The plans you create each day are the backbone of your math program.

If your school has a standard lesson plan format for its teachers, you will be expected to follow it. However, if your school allows teachers the freedom to use a format of their own, create a structure that best fits your teaching needs. Consider using or adapting the following daily lesson plan format and saving the electronic version on your computer. Write your plans on the template and save and print them out each day. Or print or photocopy the plain template, and write your plans on paper. Whichever way you choose to write your daily lesson plans, the following elements will help you create lesson plans that will satisfy the requirements of your curriculum and meet the needs of your students. Each daily lesson plan should contain the following:

- Topic, which is the content that students will be studying for this class period.
- State math standard(s), which your lesson addresses.
- Objective(s), which is what you expect your students to be able to do at the end of the lesson. Some plans may have only one objective; some may have two or three. Avoid including more objectives than you can reasonably expect your students to achieve by the end of the lesson.
- Procedure, in which you detail the steps throughout the lesson that will help your students meet your objectives. Procedures should take into account the abilities and prior knowledge of your students.
- Activities that address diverse learning styles and actively involve students in learning.
- Resources and materials that are essential for the lesson's success.
- Closure, which you may think of as the last big thought that ties the various aspects of the lesson together for your students. Your closure should contain the essential information you want students to remember from the lesson.

- ⊙ Assessment, which is how you will evaluate student achievement.
- ⊙ Assignment, which is based on the topic of the lesson and which serves as a means of reinforcing what has been taught.

You should keep all of your daily lesson plans on file. If you are teaching the same course next year, you can revise and improve your plans. In some cases, minor updating or revision can result in improved plans. Always evaluate the effectiveness of your lesson plans. It is unrealistic to expect that all of your plans will work as you intended. On the back of each plan that is written on paper (writing on the back of paper copies keeps them clean for photocopying should an administrator wish to see them later), or at the bottom of plans stored as an electronic file, write a few notes regarding how well students responded to the plan and how well they achieved your objectives for the lesson. Including suggestions for how you might make the lesson more effective enables you to improve it in the future.

Daily Lesson Plan Format

Teacher: _____ Course: _____ Period: _____

Unit: _____ Lesson: _____ Date: _____

I. Topic

II. State Math Standard(s)

III. Objective(s)

IV. Procedure

Daily Lesson Plan Format (continued)

V. Activities

VI. Resources and Materials

VII. Closure

VIII. Assessment

IX. Assignment

NOTES:

For an example of a lesson plan, see "Sample Daily Lesson Plan" that follows. Note that the sample lesson is a part of a unit on rational numbers, which served as the example for the sample unit plan provided in "Unit Plans," previously in this section.

When planning any lesson, always keep in mind the needs and abilities of your students. Plans that address specific, measurable objectives, meet course requirements, and satisfy the learning styles of your students will make your classroom an engaging center for learning math.

Sample Daily Lesson Plan

Teacher: _Mrs. Sanchez_   Course: _7th Grade Math_   Period: _3 and 4_

Unit: _Rational Numbers_   Lesson: _Adding Integers_   Date: _2/5/10_

I. Topic

Adding Integers

II. State Math Standard

Use and explain procedures for performing calculations with integers and rational numbers (pencil and paper, mental math, and calculator).

III. Objective

To develop and use a counter model to represent addition of integers.

IV. Procedure

- _Do-now: Express each of these situations as an integer: a gain of 3 yards, 7 feet below sea level, 2 strokes below par, winning 10 points, paying your brother 5 dollars._

- _Explain to the class that you can model the addition of integers using two-color counters. The red counters represent negative numbers; the yellow counters represent positive numbers. Use red and yellow counters designed for the overhead projector to model addition of integers. Place two yellow counters and three yellow counters on the overhead. Ask students to follow along, write a number sentence, and find the sum. Ask students to provide their answers; write the equation on the transparency. Follow the same procedure using the red counters. Then place one yellow counter and one red counter on the overhead and ask students to write a number sentence and find the sum. Ask them to explain how they know the sum is zero and how zero is represented by using the counters. Present additional problems that require students to model addition using positive and negative integers._

Sample Daily Lesson Plan (continued)

V. Activities

- *Students will break into small groups. Distribute two-color counters for each group. Write number sentences on the overhead for students to model in their groups and find the sums. Circulate and check students' work. Discuss the answers.*
- *Go to www.nlvm.usu.edu/en/nav/frames_asid_161_g_2_t_1.html to model addition of integers using virtual manipulatives. Tell students to write five problems they completed on the computer.*

VI. Resources and Materials

Two-color counters for students; red and yellow counters designed for use with the overhead projector; blank transparency; markers; computers with Internet access.

VII. Closure

Ask students to write a rule they can use to add integers.

VIII. Assessment

Check students' group work, their answers, and review of homework.

IX. Assignment

Text page 243, #22–35, 37–38.

NOTES:

Lesson achieved objectives. Students especially enjoyed working with virtual manipulatives.

Overcoming Common Problems in Planning

As you work to plan interesting and effective math lessons for your students, you must be aware of factors that can disrupt even the best plans. When planning, you must give close attention to pacing, pull-out programs, and special days and school activities. You must be ready for the unexpected and be able to adjust smoothly.

PACING

Proper pacing is essential for planning. Moving too quickly can make it difficult for students to master material; moving too slowly risks boring students, which can lead to disruptive behavior and problems with classroom management. You need to pace your lessons so that you can complete the required course material and all of your students can meet the objectives of your curriculum.

Following are some suggestions for proper pacing:

- Know your students. Once you understand their strengths and weaknesses in math, you are better able to gauge how quickly they can learn new concepts and skills. You can then pace lessons appropriately.

- Make the most of your class time. Whether you teach in forty-five-minute periods or eighty-minute blocks, plan lessons and activities that engage students in meaningful work for the entire class.

- Use sponge activities to fill brief gaps and transitions between classroom activities. Sponge activities, which are only a few minutes long, can be math exercises that keep your students focused on math, encourage productive behavior, and maximize time for learning. They are ideal for students who complete classwork early.

- Use anchor activities to provide students with meaningful work when they are not actively engaged in classroom activities. An anchor activity, which is directly related to the curriculum, can be worked on throughout a unit.

- Be flexible with your plans. On some days, everything in a lesson works wonderfully. On others, however, the same activities in the same lesson move ahead slowly in fitful starts and stops. Avoid rushing ahead simply to keep on your schedule. If necessary, take extra time for students to meet your objectives, which is more important than keeping to time limits.

- Always have extra work ready. There will be days, especially during group activities, when some students finish sooner than others. Providing extra work—perhaps sponge or anchor activities, reviews, or enrichment—keeps early finishers working while other students complete classwork.

⊙ Provide meaningful homework assignments. Homework that reinforces or expands what students learned in class is a great tool for maximizing learning.

Proper pacing is necessary for transforming your plans into effective instruction, ensuring that students have the right amount of time to achieve the objectives of your lessons. Without proper pacing, your math lessons—no matter how interesting—will run either too short or too long and will be disappointing to you and your students.

PULL-OUT PROGRAMS

If your school is like most, you must cope with pull-out programs in which students leave your class. Typical pull-out programs include basic skills, chorus, and band or instrument lessons.

Although these programs are beneficial for students, they can cause planning headaches for classroom teachers. Imagine the student who is an excellent musician but who struggles in math, and his trombone lesson pulls him from your math class every Monday morning. Or consider when the chorus rehearsal for the fall concert is scheduled during your geometry class and eight chorus members will not be in your class for a test on the properties of triangles. Planning is unquestionably more difficult when students are in and out of class.

To manage the potential disruption that pull-out programs can cause, establish routines for students who are pulled from your class. Once established, be consistent with the routines. For example, if a student must leave your fourth-period math class every Thursday morning, ask him to see you right after school to obtain his assignment. If possible, have several students who missed class meet with you at the same time. You may be able to present a minilesson highlighting the major concepts of the classwork they missed. Receiving their work the same day it is assigned makes it easier for these students to keep up with the class.

In addition to knowing which of your students will be regularly pulled from your classes, be aware of special occasions that will further affect pull-outs. If the drama club is rehearsing for a play during your sixth-period pre-algebra class, and several of your students will miss class, you must plan accordingly. Instead of presenting a new, somewhat challenging lesson on solving equations, plan a review for the day. You should also maintain clear communication with teachers who are in charge of pulling students out of your class, and be aware of any announcements regarding pull-outs. This will help you to adjust plans as needed to accommodate these students.

Pull-out programs can be annoying to classroom teachers. Some teachers try to ignore them, but this only results in placing burdens on you and your students for making up their missed work. A better strategy is to recognize that pull-out programs are a part of your school day and create plans that minimize the distractions they may cause.

LESSONS AND ACTIVITIES FOR SPECIAL DAYS

Every school year has several days in which regular schedules are not followed. Days before breaks, holidays, or special celebrations such as Multicultural Day, Field Day, or the day before Homecoming often require special planning.

On these days you must decide what is best for your students. Starting a new unit on the day before Homecoming when students are thinking about a pep rally, parade, football game, and dance will likely not result in many students achieving your lesson's objectives and thus require you to reteach material. Planning a lesson that focuses on review or enrichment might be a better use of class time.

Some days and months lend themselves to special math events. These events can increase student awareness and interest in math, and be easily integrated into your plans to provide a change of pace from your daily routines. Ideas, activities, and teaching suggestions for the special days and months listed below may be found at the Web sites provided. For additional information and other Web sites, search the Internet using the name of the particular event.

- ◎ October—National Metric Week is an annual event held during the tenth month of the year in the week containing the tenth day. Sponsored by the National Council of Teachers of Mathematics (NCTM), its mission is to promote the importance and convenience of using the metric system. (www.nctm.org)

- ◎ November—American Education Week is sponsored by the National Education Association (NEA) and is usually held the week preceding Thanksgiving. The goal of American Education Week is to show parents, teachers, and students the importance of education. (www.nea.org/grants/19823.htm)

- ◎ March 2nd—Read Across America Day is designed to promote student interest in reading. Initiated by the NEA, it is held annually on Dr. Seuss's birthday and is a great time to encourage students to read math literature. (See "Incorporating Literature into Your Math Class" later in this section.) (www.nea.org/grants/886.htm)

- ◎ March 14th—*Pi* Day is celebrated on the third month, fourteenth day of the year (thus 3.14) and celebrates the wonders and applications of *pi*. (www.piday.org)

- ◎ April—April is Math Awareness month, the goal of which is to increase public understanding and appreciation of mathematics. (www.mathaware.org)

- April—Although the specific date varies from year to year, Teach Children to Save Day is held every April. Sponsored by the American Bankers Association, it is designed to teach students the importance of saving money. (www.aba.com)

When planning lessons for special days or events, avoid simply providing students with random worksheets that have little relevance to your curriculum. Instead make sure that the lesson's objectives support the goals of your overall math program.

ADJUSTING LESSON PLANS

Whenever planning, always keep in mind that lessons do not always go "according to plan." There will be times that students cannot stay on task (for no apparent reason), when they are unable to meet your objectives, or are frustrated by the work. Maybe equipment malfunctions in the middle of a lesson, or a fire drill occurs just as students have settled into groups. Perhaps a problem arises between two students that you must address immediately.

For these and similar situations, you must be flexible. Always have an alternative plan ready that will enable students to achieve your original objectives for the lesson.

Following are some suggestions for changing plans during class:

- Switch to a different method of instruction or learning style. If students are finding it difficult to visualize rotating a right triangle $90°$ clockwise using Geometer's Sketchpad, provide a model of the triangle and a grid so that students can actually rotate the triangle $90°$.

- Be prepared to teach a concept that you thought students had mastered. If you are planning to use the point-slope form of an equation in your trigonometry class and students are unfamiliar with it, you must be able to switch from your plans and provide the prerequisite instruction.

- If you intend to use the Internet to gather data, have available the same (or similar) data in another form as backup. Should your Internet connection go down, you will be able to continue instruction.

- Organize students into small groups when they are having trouble understanding new concepts. Reteach the major ideas and present activities in which students can work cooperatively to explore concepts.

- Permit students to work with a partner. This can be especially helpful when working individually proves to be difficult.

- Adjust the amount of work you assign. If you intended for students to do ten example problems during class and ten more problems for homework, but students had numerous questions during your explanation which lasted the entire period, assign fewer problems for students to complete at home. If, however, students understand your explanations and examples quickly, you may wish to assign more challenging problems for homework.

The purpose of planning is to develop lessons that will help your students achieve specific objectives. Even though a lesson plan may be well conceived, sometimes, because of unanticipated factors or conditions, you must adjust plans to ensure that your students learn what you intended. Being able to modify plans quickly can result in a successful class despite such changes.

AVOIDING PLANNING PITFALLS

By recognizing the problems that can undermine planning, you will be better able to create lesson plans that will give your students interesting and successful learning opportunities.

Here are some of the most common planning pitfalls:

- Not getting to know the strengths, weaknesses, and learning styles of your students
- Not basing lesson plans on state standards, district goals, and course objectives
- Not taking into account the prior knowledge of your students as the starting point for planning
- Not planning for the diverse learning styles in your classes
- Not including activities that address critical thinking
- Not establishing a backup plan in the event of technology glitches
- Not maintaining realistic and proper pacing
- Not adjusting lessons as needed
- Not including problems that relate to students' lives
- Not including problems that connect math to other subject areas
- Not making the necessary accommodations for students in pull-out programs
- Not creating lessons that utilize the full amount of class time
- Not providing the necessary amount of practice and reinforcement for the mastery of new math skills and concepts
- Not providing adequate means of assessment
- Not providing closure for lessons

All good math teachers understand the importance of creating effective lesson plans that will result in the achievement of their students. The best math teachers also understand the many factors and situations that can undermine even the most effective plans. These teachers are able to adjust their plans to ensure that their students meet the objectives of the material presented.

Incorporating Literature into Your Math Class

In recent years math teachers have increasingly incorporated literature into their programs. The benefits for students are significant. Mathematics is a central part of the human experience, being found in just about everything. Try thinking of something that cannot be described, quantified, or validated with mathematics. Literature offers teachers a unique way to share this understanding of math with their students.

Many books are available that can help students comprehend abstract mathematical concepts, gain greater understanding of mathematical relationships, and realize how important math is to our lives, cultures, and world. Various books, including fiction, nonfiction, anthologies, short stories, and biographies, can pique student interest and lead to greater understanding and appreciation of mathematics.

There are many ways you can incorporate literature in your math class to supplement your curriculum, including:

- ◎ Assign or read books about topics that are presented in a fictionalized yet accurate manner. For example, *What's Your Angle, Pythagoras? A Math Adventure* (elementary/middle school) by Julie Ellis shows how Pythagoras discovered the theorem named for him, along with a clear, detailed explanation of the theorem. If your students are learning about circles, ask them to read *Sir Cumference and the Dragon of Pi: A Math Adventure* (elementary/middle school) by Cindy Neuschwander.

- ◎ Incorporate historical material relating to your current math topic. If your students are about to study complex numbers, read an excerpt about how this concept evolved in Sanderson M. Smith's *Agnesi to Zeno: Over 100 Vignettes from the History of Math* (middle school/high school).

- ◎ Use biographies of mathematicians found in such books as *Mathematicians Are People, Too, Volumes I and II* (middle school/high school) by Luetta and Wilbert Reimer to learn about the lives of some people who contributed to the development of mathematics.

- ◎ Use books to explore math and cultural themes. For example, *Africa Counts: Number and Patterns in African Culture* (high school) by Claudia Zaslavsky explores how African cultures work with numbers and patterns.

- ◎ Collaborate with other teachers. A unit on the pioneers can be enhanced in language arts, history, and math by reading *Sweet Clara and the Freedom Quilt* (elementary/middle school) by Deborah Hopkinson.

- Read excerpts from books such as *Math Mini Mysteries* (middle school/high school) by Sandra Markle to develop problem-solving strategies and reasoning abilities.

- Use books to illustrate math concepts that are difficult to visualize. *How Big Is It?* (elementary/middle school) by Ben Hillman is a wonderful way to help your students to visualize relative sizes.

- Recommend books that are geared to the interests of individual students. For example, your most motivated students may enjoy *Career Ideas for Kids Who Like Math* (middle school/high school) by Diane Lindsey Reeves. Students who are Star Trek fans will be fascinated by *Computers of Star Trek* (high school) by Lois Gresh and Robert Weinberg.

Following are more books you may consider. For your convenience, the books are broken down into three general grade-level categories: Elementary/Middle School, Middle School/High School, and High School.

ELEMENTARY/MIDDLE SCHOOL

- Adler, David A. *How Tall, How Short, How Far Away?*
- Allison, Linda, Marilyn Burns, and David Weitzman. *The I Hate Mathematics Book*
- Anno, Mitsumasa. *Anno's Math Games*
- Burns, Marilyn. *Math for Smarty Pants*
- Calvert, Pam. *Multiplying Menace: The Revenge of Rumpelstiltskin (A Math Adventure)*
- Clement, Rod. *Counting on Frank*
- Enzensberger, Hans Magnus. *The Number Devil: A Mathematical Adventure*
- Lasky, Kathryn. *The Librarian Who Measured the Earth*
- McCallum, Ann. *Rabbits Rabbits Everywhere: A Fibonacci Tale*
- Nagda, Ann Whitehead, and Cindy Bickel. *Tiger Math: Learning to Graph from a Baby Tiger*
- Neuschwander, Cindy. *Sir Cumference and the First Round Table: A Math Adventure*
- Neuschwander, Cindy. *Sir Cumference and the Great Knight of Angleland: A Math Adventure*
- Neuschwander, Cindy. *Sir Cumference and the Isle of Immeter: A Math Adventure*
- Neuschwander, Cindy. *Sir Cumference and the Sword in the Cone: A Math Adventure*

- Packard, Edward. *Big Numbers: And Pictures That Show Just How Big They Are!*
- Pittman, Helena Clare. *A Grain of Rice*
- Schwartz, David M. *How Much Is a Million?*
- Scieszka, Jon. *Math Curse*
- Tang, Greg. *The Grapes of Math: Mind-Stretching Math Riddles*
- Tompert, Ann. *Grandfather Tang's Story*
- Wells, Robert E. *Can You Count to a Googol?*

MIDDLE SCHOOL/HIGH SCHOOL

- Blatner, David. *The Joy of Pi*
- Bruce, Colin. *Conned Again, Watson: Cautionary Tales of Logic, Math, and Probability*
- Burger, Dionys. *Sphereland: A Fantasy About Curved Spaces and an Expanding Universe*
- Cooney, Miriam P. *Celebrating Women in Mathematics and Science*
- Dewdney, A. K. *A Mathematical Mystery Tour: Discovering the Truth and Beauty of the Cosmos*
- Dewdney, A. K. *200% of Nothing: An Eye-Opening Tour Through the Twists and Turns of Math Abuse and Innumeracy*
- Field, Robert. *Geometric Patterns from Roman Mosaics: And How to Draw Them*
- Frucht, William. *Imaginary Numbers: An Anthology of Marvelous Mathematical Stories, Diversions, Poems, and Musings*
- Henderson, Harry. *Modern Mathematicians*
- Isdell, Wendy. *A Gebra Named Al: A Novel*
- Johnson, Art. *Classic Math History Topics for the Classroom*
- Juster, Norton. *The Dot and the Line: A Romance in Lower Mathematics*
- McKellar, Danica. *Kiss My Math: Showing Pre-Algebra Who's Boss*
- Pappas, Theoni. *Fractals, Googols and Other Mathematical Tales*
- Pappas, Theoni. *Math Talk: Mathematical Ideas in Poems for Two Voices*
- Pappas, Theoni. *Mathematical Footprints: Discovering Mathematical Impressions All Around Us*
- Paulos, John Allen. *Once Upon a Number: The Hidden Mathematical Logic of Stories*
- Perl, Teri. *Math Equals: Biographies of Women Mathematicians + Related Activities*
- Sachar, Louis. *More Sideways Arithmetic from Wayside School*

- Sandberg, Carl, and Ted Rand. *Arithmetic: Illustrated as an Anamorphic Adventure*
- Schimmel, Annemarie. *The Mystery of Numbers*
- Schwartz, David M. *If You Hopped Like a Frog*
- Schwartz, David, M. *On Beyond a Million: An Amazing Math Journey*
- Stein, Sherman K. *How the Other Half Thinks: Adventures in Mathematical Reasoning*
- Stein, Sherman K. *Strength in Numbers: Discovering the Joy and Power of Mathematics in Everyday Life*
- Tahan, Malba. *The Man Who Counted: A Collection of Mathematical Adventures*

HIGH SCHOOL

- Abbott, Edwin A., and Ian Stewart. *Flatland: A Romance of Many Dimensions*
- Aczel, Amir D. *Descartes's Secret Notebook: A True Tale of Mathematics, Mysticism, and the Quest to Understand the Universe*
- Aczel, Amir D. *Fermat's Last Theorem*
- Beckmann, Peter. *A History of Pi*
- Berlinski, David. *A Tour of Calculus*
- Bodanis, David. $E = mc^2$ *: A Biography of the World's Most Famous Equation*
- Cajori, Florian. *A History of Mathematical Notations*
- Dunham, William. *The Mathematical Universe: An Alphabetical Journey Through the Great Proofs, Problems, and Personalities*
- Eastaway, Rob, and Jeremy Wyndham. *Why Do Buses Come in Threes? The Hidden Mathematics of Everyday Life*
- Fadiman, Clifton. *The Mathematical Magpie*
- Flannery, Sarah, and David Flannery. *In Code: A Mathematical Journey*
- Gillings, Richard. *Mathematics in the Time of the Pharaohs*
- Guillen, Michael. *Five Equations That Changed the World*
- Hoffman, Paul. *Archimedes' Revenge: The Joys and Perils of Mathematics*
- Hoffman, Paul. *The Man Who Loved Only Numbers: The Story of Paul Erdos and the Search for Mathematical Truth*
- Humez, Alexander. *Zero to Lazy Eight: The Romance of Numbers*
- Kaplan, Robert. *The Nothing That Is*
- Maor, Eli. *e: The Story of a Number*
- Nasar, Sylvia. *A Beautiful Mind: The Life of Mathematical Genius and Nobel Laureate John Nash*

- Paulos, John Allen. *A Mathematician Reads the Newspaper*
- Peterson, Ivars. *The Mathematical Tourist: Snapshots of Modern Mathematics*
- Salsburg, David. *The Lady Tasting Tea: How Statistics Revolutionized Science in the Twentieth Century*
- Seife, Charles. *Zero: The Biography of a Dangerous Idea*
- Singh, Simon. *Fermat's Enigma*
- Zebrowski, Ernest. *A History of a Circle*

You can find more information about literature and math by searching the Internet with "literature and math," "incorporating literature and math," or similar terms. When you incorporate literature, you expand your math program, spark your students' interest, and provide them with greater opportunities for learning and gaining an awareness of mathematics.

Incorporating Writing into Your Math Class

The ability to write clearly is a valuable skill. Writing should not be taught only in language arts and English classes, but should be a part of every subject. It is a particularly important part of a math program.

In order to write effectively, students must gather, analyze, and organize ideas. To communicate their ideas, they must compose their thoughts so that they are clear and logical. Through writing, students are able to explore concepts in math, discover relationships, and connect math to other subjects. Writing enables students to go beyond basic understanding to a deeper and broader comprehension of mathematics.

As many students may not feel comfortable writing in math class, you must explain the value of writing in your class. Discuss how writing helps to clarify their thinking and improves their skills in communication. Give them examples of what you consider to be good writing—for example, a description of how to solve a specific math problem.

You have several options for integrating writing into your math program. The most common include:

- Math journals (See "Math Journals" and "Math Journal Writing Prompts" in Section Seven)
- Explanations of solutions to specific problems
- Conjectures about new concepts and ways to test them
- Answering questions in which students reflect on what they learned in a lesson or unit
- Summaries or explanations as a part of math projects

- Reports on topics in math
- Articles for publication in a class math newsletter or magazine
- Postings regarding math problems, news, or information on your school or class Web site

Whenever your students write formally about math, encourage them to follow the stages of the writing process, which they have most likely studied in their English or language arts class. The writing process describes the way people write effectively. It has five important stages: prewriting, drafting, revising, editing, and publishing (or sharing). You may find it helpful to hand out copies of "The Writing Process," which follows, and discuss the stages with your students.

The Writing Process

Name _____ Date _____ Period _____

When you write, be sure to use the stages of the writing process.

- **Prewriting:** Think of a purpose for writing. Gather, generate, analyze, and organize ideas.

- **Drafting:** Set your ideas in written form.

- **Revising:** Refine, delete, or add ideas to your draft. You may rewrite the draft several times.

- **Editing:** Proofread and make any final changes to your writing.

- **Publishing (or sharing):** Communicate your ideas to others.

Note that in some cases—for example, a brief explanation of the solution to a problem—students will probably not use all the stages of the writing process. However, always encourage them to formulate their thoughts and be sure they have answered any questions clearly and accurately.

Insist on good writing. Encourage your students to write clearly, explain mathematical reasoning, use precise math vocabulary, and demonstrate connections and relationships between concepts. They should build their writing around main ideas, support those ideas with details, and use complete sentences, paragraphs, and correct capitalization and punctuation. Consider consulting with your students' English teacher about the writing process. She may be able to help your students as they write in your math class.

Respond to the writing of your students. Rather than simply correcting their papers, write notes of encouragement, suggestions how they might clarify their thoughts, or comments on their ideas, insights, or those special eureka moments.

Writing expands the boundaries of your math class. It gives students a chance to reflect on the math skills and concepts they are learning, leading to deeper understanding. No less important, it gives them a chance to examine their thoughts, feelings, and attitudes about math.

Quick Review for Designing Effective Math Lessons and Activities

Planning is a cornerstone of the foundation of a successful math program. Without effective lessons and activities, it is impossible to present a course's concepts and skills in a manner that supports and fosters learning. Effective planning is essential for effective instruction.

The following points will help you to design math lessons and activities that will result in a productive year for you and your students:

- ◎ Build time for planning into your schedule. Having time set aside for planning makes it easier to plan consistently.

- ◎ Always have all of the necessary materials and information handy when you plan.

- ◎ Plan a variety of interesting lessons and activities that address the needs of your students.

- ◎ Base specific objectives in your plans on your curriculum, state standards, and district goals.

- ◎ Create effective lessons and activities by:
  - • Addressing the needs of diverse learners
  - • Basing the presentation of new material on your students' prior knowledge and abilities

- Selecting activities and problems that are relevant to students' lives
- Incorporating questions and activities that stimulate critical thinking
- Incorporating the use of technology whenever possible
- Including closure
- Providing a means of assessment

○ Divide your curriculum into unit plans that address major topics.

○ Divide unit plans into daily lesson plans.

○ Be aware of common problems in planning, and work to overcome them, particularly lack of proper pacing, not utilizing the full amount of time in class, and not having backup plans for the unexpected.

○ Incorporate literature in your math program.

○ Incorporate writing in your program.

Creating lessons and activities that are based on clear objectives, facilitate learning, and foster your students' interest in math is one of your most important and ongoing responsibilities. It is through unit and daily lesson plans that you transform the extensive material of your math course into manageable parts that you can teach and that your students can understand and master.

SECTION ELEVEN

Providing Effective Math Instruction

Without thorough, engaging, and interesting lesson plans, a math teacher cannot provide his or her students with effective instruction. The converse is also true, however. Unless a teacher is able to deliver effective instruction, the best lesson plans will result in minimal learning.

Effective instruction is based on various factors, including understanding the complex and demanding role of math teachers today; using a variety of instructional methods; tailoring instruction to meet the needs of diverse learning styles; managing the classroom to optimize learning; and involving students in the learning process by encouraging them to be active rather than passive learners. Delivering effective instruction requires that you hone your teaching skills, and develop a teacher "presence" that enables you to gain the respect and attention of your students, and simultaneously inspire them to want to learn mathematics in your class.

Math teachers who plan great lessons and then efficiently and effectively deliver those lessons to students acquire a reputation as being among the best math teachers in their school. Students like being in their class because the class is enjoyable, purposeful, and successful.

Being a Facilitator of Learning

Not too long ago, the typical math teacher conducted class by standing at the front of the classroom where he explained math concepts and skills and wrote notes and examples on the chalkboard. After the lesson's explanation, he would have assigned practice problems for students to work on in class, and then assign more, similar problems for homework. There was little use of technology, minimal emphasis on creative problem-solving, and infrequent, if any, sharing of ideas. It was scant wonder that few students, other than those who possessed an innate ability and interest in numbers, looked forward to math class.

Math instruction today involves much more than simply introducing new material and assigning problems for students to solve. Math teachers must assume many roles if they are to provide effective instruction to their students.

During any class, your roles may switch among the following:

Role Model	Coach
Planner	Empathizer
Problem solver	Critical thinker
Organizer	Decision maker
Encourager	Questioner
Cheerleader	Surrogate parent
Mentor	Disciplinarian
Listener	Speaker
Evaluator	Demonstrator
Helper	Counselor

Of course, above all these roles is the role of teacher. As you assume your numerous roles, you must remember that you are in charge of the class and responsible for student learning. One of the roles you cannot assume is that of being a friend to students. You must maintain professional distance, yet be involved enough with students to help them realize their potential.

Meeting the Needs of Diverse Learners Through Instruction

All of your students are individuals. They have separate needs, abilities, and learning styles. In "Addressing the Needs of Diverse Learners" in Section Ten, we reviewed basic learning styles: the auditory learner, who acquires information best by hearing; the verbal visual learner, who acquires information best through written language; the nonverbal visual learner, who acquires information best through visual displays such as charts; and the kinesthetic learner, who acquires information best through touch and movement. Although students may learn best through one particular learning style, most are able to learn through various styles and all students benefit from learning through a variety of methods.

Personality also plays a part in learning. Some students are enthusiastic, independent learners. Others are dependent learners, and some are, unfortunately, unmotivated and disinterested. Some prefer to work alone, and others thrive in groups. Together, personality and learning styles make for a dizzying assortment of factors you must consider when attempting to meet the individual learning needs of your students and help them become successful in your class.

The best strategy is to view all of your students as individuals with unique strengths, weaknesses, and abilities. Once you recognize the needs of each student, you can modify your teaching to satisfy those needs. For more information and a more detailed breakdown on learning styles, visit www.learningstyles-online.com.

Following are tips for identifying the basic learning styles of your students:

- Note which students prefer lectures and explanations. These are likely to be auditory learners.

- Note which students prefer to read material. These are likely to be verbal visual learners.

- Note which students prefer to work with pictures, drawings, graphs, illustrations, and charts. These students are likely to be nonverbal visual learners.

- Note which students prefer to work with models and manipulate objects. These are likely to be kinesthetic learners.

To reach all the different types of learners in your class, you must provide information in a variety of ways. Only by recognizing and addressing the learning styles of your students can you guide them to success in math.

Using Various Instructional Methods for Teaching

Because your students have a variety of preferred learning styles, you should try to use a variety of instructional approaches and activities in your lessons. Only by varying instruction can you meet the learning needs of all your students.

Following are several instructional methods common to math classes, the learning styles they most benefit, and an example of how the method might be used in a math class:

- *Lecture and examples:* Auditory learners. You can use lectures and examples to convey complicated algorithms and procedures that require detailed explanations.

- *Questions and answers:* Auditory learners. Question-and-answer sessions help students to recall knowledge and apply it to new skills and concepts.

- *Demonstration:* Auditory and nonverbal visual learners. You can demonstrate a reflection on a coordinate plane by taking a figure and showing its reflection in a given line.

- *Presentation:* Auditory, verbal and nonverbal visual learners. A PowerPoint presentation on functions is an excellent method for highlighting the properties of different types of functions.

○ *Investigation:* Kinesthetic learners. You can ask students to measure the height of classroom items rather than your giving them the measurements.

○ *Student presentation and explanation of their work on a traditional board or interactive whiteboard:* Auditory, verbal and nonverbal visual, and kinesthetic learners, depending on the specific work. You might ask students to write and explain the solution to a homework problem.

○ *Cooperative group work:* Auditory, verbal and nonverbal visual, and kinesthetic learners, depending on the assignment. Discussion during the activity benefits auditory learners. Using manipulatives, such as pattern blocks to investigate geometric shapes, primarily addresses nonverbal visual and kinesthetic learning styles, whereas working on a report addresses verbal visual learners.

○ *Reading aloud to students:* Auditory learners. Reading a biographical sketch of René Descartes can provide interesting background for students learning about the coordinate plane. (See "Incorporating Literature into Your Math Class" in Section Ten.)

○ *Technology:* Auditory, verbal and nonverbal visual, and kinesthetic learners, depending on the activity. Gathering information from a math Web site benefits both verbal and nonverbal visual learners; discussing the information benefits auditory learners. Using calculators addresses the needs of kinesthetic learners.

○ *Visual aids:* Auditory and nonverbal visual learners. For example, showing students a picture of Pascal's Triangle to illustrate its properties addresses nonverbal visual learners. If you use the picture to explain or ask students to explain the properties of the triangle, you are targeting auditory learners as well.

○ *Graphic organizers:* Verbal and nonverbal visual learners. Composed of diagrams and words, graphic organizers help students organize, process, and retain information. For example, students can complete a graphic organizer to classify quadrilaterals.

○ *Reviews:* Auditory, verbal and nonverbal visual learners. Review-related discussions primarily benefit auditory learners, and using tables to detail information could benefit both verbal and nonverbal visual learners, depending upon the content of the table.

○ *Guest speakers:* Auditory learners. Because they talk about specific topics in their area of expertise and promote discussion, guest speakers benefit auditory learners.

○ *Field trips:* Auditory, verbal and nonverbal visual learners, and kinesthetic learners. Depending on the destination and activities of a field trip, all learners can benefit.

By employing a variety of instructional methods and activities, you can stimulate student interest, keep your math lessons fresh and interesting, and address your students' different learning styles. You will make learning easy for your students.

Fostering Problem-Solving Skills in Your Students

Helping your students develop problem-solving skills should be a priority in your math program. An ability to analyze complex problems, decide on the best procedure for finding solutions, and working toward solutions in a methodical manner requires skills that students will use throughout their lives.

To promote the development of problem-solving skills in your students, you should:

- Provide students with challenging problems. Some problems should have multiple steps and several possible solutions.

- Provide students with problems that are relevant to their lives and that connect math to other subjects.

- Encourage your students to be open-minded and use various strategies to solve problems.

- Require students to explain their solutions.

- Use a rubric that tells students how their solution to a problem will be assessed. This can provide guidance for solving the problem. (See "Assessment Through Open-Ended Problems" in Section Twelve.)

- Include time for sharing the results of problem solving that highlight the use of different strategies.

- Allow time for students to reflect on the procedures they used and solutions they found in their own work and that of their peers.

To further help your students develop problem-solving skills, distribute and discuss the following "Steps for Solving Math Word Problems." Suggest that your students keep the sheet in their binders and refer to it periodically, or as necessary.

When you teach your students problem-solving skills, you are giving them skills that they can apply to problems in many different situations. Problem-solving skills are practical skills that help students to become successful learners.

Steps for Solving Math Word Problems

Name _____ Date _____ Period _____

Follow the steps below to solve those tough math word problems.

1. Read and study the problem. Determine what the question is asking.

2. Find the information you need to solve the problem. Eliminate any unnecessary information. Supply missing facts.

3. To help you understand the problem, consider doing the following:

   - Write equations.

   - Organize data by making tables and charts.

   - Sketch or draw a model, graph, or diagram.

   - Use trial and error (also called guess and check).

   - Find patterns and relationships.

   - Think about the problem from a different point of view.

   - Work backward.

   - Use logical reasoning.

   - Simplify the numbers by rounding or estimating.

   - Think about a similar problem you have solved and use similar strategies.

   - Consider using a calculator to investigate tables and graphs.

   - Use multiple strategies.

4. Write notes as you work toward a solution. Refer back to your notes as necessary to keep moving forward.

5. Decide what operation or operations you will need to solve the problem. Perform the operations.

6. Double-check your work. Make sure it is accurate and that your answer is reasonable. Be certain that you answered the entire question.

Motivating Your Students

Students who are motivated experience greater achievement than those who are unmotivated. Although some students are self-motivated, most students need some extra motivation and encouragement if they are to attain their potential. Effective teachers know how to inspire their students to strive for success.

The following list contains tips for motivating your students:

- Begin each lesson with a hook—something that grabs the attention of your students immediately. The hook might be a puzzling statement, the results of a survey, or a surprising statistic. The hook should tie directly into the lesson. For example, for a lesson introducing integers you might ask students: How can something be less than zero?

- Weave the content of the lesson into students' lives. For example, if many of your students enjoy sports, tie the subject into sports. For a lesson about fraction and decimal equivalencies, explain that batting averages in baseball are expressed in decimal form. A batter who has one hit in three at bats has a batting average of .333. (Baseball batting averages are written using three decimal places, without a zero to the left of the decimal point.)

- Maintain high but realistic expectations for your students. If students feel that you do not believe they can complete challenging work, they may not be willing to try to solve hard problems.

- Illustrate concepts in an unusual manner. For example, have students indicate the slopes of lines by extending their arms.

- Use activities that require creative, open-ended thinking. A challenge always generates interest. For example, asking students to create a schedule for after-school clubs is a great application of discrete math.

- Connect material to the practicalities in students' lives. Show how what they are learning is beneficial. If you are teaching discounts and sale price, share with students sales announcements that advertise items reduced by percentages.

- Involve students with the lesson. Ask students questions, and encourage them to ask you questions. Try to call on as many students as possible each day, and make certain that you call on every student regularly.

- Assign cooperative activities when possible. Most students enjoy working in groups. Tasks that might seem overwhelming alone can be manageable in a group setting.

- Connect skills that students learn in math to other subjects. For example, learning about and using the metric system in math can help students apply metric measurements in science class.

- Display motivational posters in class. Such posters can promote a can-do atmosphere in class and encourage students to always work hard.

- Use positive reinforcement. Offer genuine compliments (students can always see through false ones) that address specific behaviors, write words of encouragement or praise on tests and quizzes, and respond to student writing in journals and on reports. Use stickers, which are available in office supply stores, to reward success. (Even high school students like receiving stickers acknowledging their efforts.) Everybody likes to hear that they are doing a good job and moving forward. A few short words of encouragement can go a long way toward motivating the unmotivated student.

Let your students know that you believe that they can do well in your class, and they will believe in themselves. Motivated students remain focused during instruction, and work hard both in and outside of school.

Improving Your Instructional Delivery Skills

There is much validity in the observation that teaching, and particularly delivering instruction, is similar to "putting on a show." Math teachers who are able to make mathematics exciting and meaningful for students are usually the teachers whose students are the most successful. From speaking clearly and confidently at the start of your lesson to accurately gauging the responses of your students at its end, delivering effective instruction is a result of several factors you can control.

SPEAKING

When you deliver instruction to your students, you must speak with clarity and enthusiasm. You should understand your topic fully and know exactly what you intend to say. Never forget that when teaching students, who may or may not be interested in your subject, how you say something is often as important as what you say.

Following are tips that can help you speak effectively when delivering instruction:

- Review your lesson plans before class. Being entirely familiar and comfortable with the material will enable you to present it smoothly.

- Rehearse your lesson. Know the sequence of information you intend to provide to your students, the questions you will ask them during the lesson, and the outcomes you expect. Try to anticipate questions students will ask you and be prepared to provide answers, as well as background and supplemental material if necessary.

- Speak loudly enough so that all students can hear you easily. Speak in a pleasant tone with clear enunciation. Vary your tone.

○ Speak with liveliness and confidence. Remember that you are the math expert in your class. When you convey excitement about the topic, your students are likely to become excited, too.

○ Speak the language of math. Use correct grammar and precise mathematical terms.

○ Practice to eliminate filler words such as "like," "um," and "ah" when you speak. Such words can undermine the concentration of your students and make it hard to follow the development of your ideas.

○ Emphasize and repeat, if necessary, important words and phrases.

○ Pause if you feel your students are having trouble understanding an idea or skill. Ask for a student volunteer to explain what you said in his or her own words to the class. If necessary, reexplain the material yourself.

○ Make certain that any directions or instructions you give are clear. Again, ask for volunteers to repeat the directions in their own words.

○ Make eye contact with your students as you speak. Avoid focusing on the back wall or looking away from students while you are speaking. Teach to the whole class. During the presentation of your lesson, look from student to student, meeting their eyes for a second or two. Maintaining eye contact during the lesson makes students feel that you are speaking directly to them and helps to keep their attention. This also reduces opportunities for students to misbehave.

○ Think aloud. Let students hear your thought processes as you work out a problem on the board in answer to a student's question.

○ Avoid talking to the board as you write example problems. Turn your body slightly back to the class as you write and speak. This will allow you to project your voice to the class and will also enable you to easily glance at students to make certain everyone is focused on the lesson.

When you present a lesson and utilize effective speaking skills, you not only deliver the material clearly, but you are also more likely to hold the interest of your students throughout the lesson. The manner in which you speak during your presentation has the power to excite students about the topic.

BODY LANGUAGE

When you smile, frown, or stand tense in anger, your body language (often referred to as nonverbal cues) is communicating your thoughts and feelings just as plainly as words. When your body language matches your words, you communicate clearly to your students. However, when your body language does not match your words, your students can be confused because they are receiving mixed signals.

Body language can enhance as well as undermine your delivery of instruction. Use nonverbal cues to emphasize your expression and avoid cues that seem to contradict what you are saying. Here is an example. You are teaching an essential concept to new material, and a student asks you a question. As she speaks, you turn to the board and continue writing an example. Although you always encourage students to ask you questions and you are listening to this student, your action in turning away may make her feel that you are not interested in her question. The message she receives is muddled. Your turning away may in fact make a greater impression on her than your answer to her question.

Following are some tips about using body language effectively during the delivery of instruction:

- Stand confidently before your students with your shoulders straight.
- Smile often and nod in approval or agreement.
- Look at your students when you are speaking to them and when they are speaking to you. Make eye contact.
- Lean forward slightly to show your interest when students ask or answer a question.
- Avoid negative nonverbal cues such as the following:
  - Frowning or rolling your eyes in disgust or disapproval
  - Standing rigidly with your hands on your hips or your arms crossed in front of you
  - Pointing angrily at your students
  - Standing too close to students, or leaning forward into their personal space
  - Looking away from a student when she is speaking to you
  - Appearing disinterested
  - Fidgeting when students are speaking to you

Most of us are unaware of how much information our bodies communicate. Our gestures, stances, and facial expressions can reveal our thoughts and feelings whether or not we wish to reveal them. Your students are attuned to the nonverbal cues of body language. They can quickly tell if you are angry, pleased, or bored with them just by how you appear. Learn to use the language of your body to enhance, rather than detract from, the delivery of your instruction.

USING THE TRADITIONAL BOARD, OVERHEAD PROJECTOR, OR INTERACTIVE WHITEBOARD

Perhaps more so than other teachers, math teachers use traditional chalkboards, whiteboards, overhead projectors, and interactive whiteboards for writing problems, notes, and examples to help students understand concepts and skills. Just as you urge your students to write neatly and clearly, so should you.

When writing on boards or projecting information onto screens, do the following:

- Be sure the board or screen is not blocked by any equipment and that all students have a clear view of it. (Also, be aware of students who might have trouble seeing the board because of possible vision problems. Alert the school nurse or their parents or guardians if necessary.)
- Speak as you write and draw, which assists auditory learners in processing the information.
- Write legibly and draw figures accurately.
- Use large plain letters and numbers.
- Write and draw large enough for all students to see easily.
- Leave enough space between words, problems, and figures for students to see information and relationships. Avoid clutter.
- Use several colors of chalk or markers for highlighting and adding visual interest.
- Always label units and dimensions.
- Erase cleanly.
- Write assignments in the same area of the board every day.
- If you share a room, as a courtesy to other teachers, erase the board when you leave the room after each class.

The value of placing notes, problems, and figures on a board or projecting them onto a screen is undermined if the information is unclear, cluttered, or hard for students to see or read. Effective use of a board, projector, and screen is a great way to model note-taking techniques, a fundamental study skill. Be sure to fully utilize these valuable teaching aids.

HANDLING MATH MANIPULATIVES EFFECTIVELY

Manipulatives are useful teaching aids. You can use manipulatives to model concepts, show relationships, and foster students' imagination and visual thinking. Through hands-on investigations with manipulatives, your students can gain a deeper understanding of mathematics.

Following are some suggestions for improving your instruction through the use of manipulatives:

- Make sure you know how to use the manipulative, particularly how it can demonstrate the concepts you wish to show your students.

- Practice the lesson using the manipulative ahead of time.
- Have the manipulatives you intend to use for a lesson ready as soon as the class begins.
- If students are to work with sets of manipulatives, prepare the sets before students arrive. Distribute the manipulatives efficiently.
- Allow time for students to investigate and explore the manipulative before they begin the assigned task. This helps them to become familiar with it.
- Demonstrate how to use the manipulative. Be sure that all students can see what you are doing.
- Consider utilizing the overhead projector and manipulatives designed for use on it.
- Provide students with clear guidelines for using manipulatives.
- Anticipate questions and problems that students might have. Try to prevent confusion and problems before they happen.
- Use and encourage your students to use the proper math terms as they explore properties and concepts.
- Encourage students to use their imagination, make predictions, and test conjectures when exploring math with manipulatives.
- Circulate around the room monitoring students as they work with manipulatives. Ask questions that foster self-discovery; answer questions that guide them forward but allow them to reach their own conclusions.
- Allow sufficient time for students to work. Avoid rushing the lesson.
- Plan an activity for students who finish early. Consider a sponge or anchor activity.
- At the end of the lesson, collect and store manipulatives properly.
- Clean manipulatives periodically.

Because they can be used for demonstrations or to provide hands-on exploration, manipulatives can help students, particularly nonverbal visual and kinesthetic learners, gain insights in math that might otherwise elude them. Manipulatives help make math tangible. (See "Common Math Manipulatives" in Section Three.)

USING TECHNOLOGY WITH EXPERTISE

Integrating technology with your instruction provides your students with rich and valuable experiences in math class. Whether you are using a digital projector or interactive whiteboard to present a lesson, or your students are using calculators or computers and the Internet during a lesson, technology opens doors to marvelous learning opportunities. It allows your students to explore and understand mathematics in a way that is impossible using pencil and paper. As with any method of instruction, you must fully understand the use of the technology you integrate into your lessons and implement it seamlessly with your instruction.

Here are some tips for using technology efficiently in math instruction:

- Have a clear purpose for using technology with specific lessons. Explain to your students why they are using a particular technology and how it aids in finding the solutions to problems.

- Make sure you have all of the necessary equipment ready before class begins (laptops, screens, digital projector, interactive whiteboard, speakers, extension cords, cables, and so on).

- Be familiar with the hardware. Know how to make the necessary connections for all components of a presentation. For instance, if you are demonstrating the menus of a calculator and using a presenter, be sure you set the equipment up correctly. Try to anticipate any glitches that might occur and how you can fix them quickly so that the lesson can move forward.

- Check the compatibility of all the equipment and software you are using.

- Practice using the technology before the lesson. Follow the steps students will use. Ask yourself questions, such as: Are the directions clear enough for students to understand? If students are to use a specific Web site for finding information, is the site still available? How much instruction will students need to be able to use the technology or software? Plan enough time for explanations and questions.

- Clearly explain the features of the technology that students will use during the lesson. Demonstrate what students are expected to do. Consider projecting calculator or computer displays on a screen as you provide instructions.

- Write the steps students are to follow on the board or a reproducible.

- Always have a backup plan for equipment failures that cannot be easily corrected. If you plan for students to use the Internet and the connection is down, you must have other materials available that students can use to finish the lesson.

- During the lesson, monitor students closely. If your students are to visit specific Web sites, make certain that they are in fact on the correct site and not checking e-mail or visiting sites such as Facebook or MySpace.

- Integrate the technology with the lesson in a manner that enables students to recognize its real-life applications. For example, if students are to search the Internet for information, explain that using specific search terms, for instance "right triangle" instead of "triangle," will lead them to the most useful Web sites on their topic.

- Model appropriate technology etiquette for your students. If computers are to be shut down, students should follow the proper procedures. Simply switching off the power can result in software problems later.

- If you do not have enough computers for every student in class, consider the following strategies:

  - Use the computer lab for the lesson.

  - Have students work in pairs.

  - Direct students to take turns.

  - Provide a series of deadlines so that computer time will be used efficiently, and that all students will have an opportunity to use a computer.

  - Suggest that students work on computers at home.

  - Encourage students to work on computers in the school media center or computer room during their free periods or after school.

You can use technology in several ways to enhance your instruction. Though you may use a digital projector or an interactive whiteboard for presenting a lesson in a clear, visually appealing manner, computers and calculators are likely to be the most important technology you and your students will use throughout the year.

Computers offer you numerous opportunities for instruction, including:

- Accessing the Internet and math Web sites; projecting Web sites on a screen or interactive whiteboard

- Gathering data and information from the Internet to create relevant problems for students to solve

- Creating WebQuests

- Using virtual manipulatives to demonstrate properties and concepts

- Demonstrating how to use dynamic software such as Geometer's Sketchpad (www.keypress.com) to explain geometric relationships and concepts

- Using Microsoft Excel to create spreadsheets, analyze data, and create graphs

- Creating a PowerPoint presentation

- Writing tests and worksheets using math software. Microsoft's Equations Editor, included in Microsoft Word, and MathType, about which you can learn more at www.dessci.com, are two programs that will enable you to write complex formulas and problems

- Previewing instructional math games

Students can also use computers in math class in various ways, including:

- Using software to explore mathematical concepts and properties

- Exploring virtual manipulatives

- Creating spreadsheets to organize data, create graphs, and solve problems
- Playing interactive math games to reinforce skills
- Completing a WebQuest
- Visiting Web sites devoted to math
- Researching mathematical topics
- Writing reports or completing projects

Calculators obviously play a major role in any math class. Some of the ways students can use calculators include the following:

- Computing with numbers
- Finding measures of central tendency
- Evaluating functions
- Graphing functions
- Solving equations and inequalities
- Creating data displays
- Creating scatterplots and finding the line of best fit
- Modeling exponential growth
- Creating and performing operations on matrices
- Generating arithmetic and geometric sequences
- Finding the sum of sequences
- Graphing parametric and polar equations
- Finding permutations and combinations
- Sharing data by linking calculators

Technology is an important part of every classroom, but it is especially important to math classes. When technology is integrated into a lesson in a purposeful manner, students not only learn math, but they learn the relevance of technology and how they can then apply it to other aspects of their lives. (See "Technology" and "Resources on the Internet" in Section Three.)

MANAGING INTERRUPTIONS AND GETTING BACK ON TASK

Every teacher knows the frustration (and irritation) that comes when an excellent lesson during which she has the full attention and interest of her students is interrupted. Whether the interruption is caused by a student who comes to class late, a fire drill, or a request by the school's secretary that a student be sent to the office, the effect is the same. Your students' concentration is disturbed, and you must stop teaching and attend to the cause of the interruption.

As you cannot eliminate interruptions, you must take steps to reduce the distraction they cause. Above all, you should always try to design and present interesting and meaningful math lessons. When your students are fully engaged in the lesson, it is easier for them to refocus their attention after an interruption. They will be anxious to get back to work.

Following are some additional strategies to consider to minimize the effect of interruptions:

- Follow your school's policies regarding students leaving the classroom. Some schools have clear guidelines; for example, students may be permitted to go to their lockers before and after school, between classes and during lunch, but not during classes. If students ask to leave the class for unacceptable reasons, remind them of school policy.

- Establish practical procedures for students leaving their seats and asking to leave class during instruction. You might, for example, request that students do not ask permission to go to the lavatory while you are presenting a lesson and new material. Explain that to leave at this time results in their missing essential information. (However, do not prevent students from going to the lavatory in the case of an emergency.) Never permit students to leave their seats to throw out a piece of paper or sharpen a pencil while you are teaching. Not only is this rude, but it allows students an excuse not to pay attention to the lesson. It also can distract others.

- To keep track of students who leave class, have a sign-out sheet near the door. Consider only allowing one student out of the class at a time, except in the case of a true emergency. (See "Classroom Sign-Out Sheet" in Section Two, and "Smoothly Handling Requests to Leave the Classroom" in Section Seven.)

- Use hall passes with the student's name and destination to ensure that students go where they are supposed to go. (See "Hall Passes" in Section Two, and "Smoothly Handling Requests to Leave the Classroom" in Section Seven.)

When you view interruptions as a normal part of your day and manage them with composure and ease, your students will be less distracted than if you show obvious annoyance at the disruption. If you are interrupted while you are teaching, simply say to your class, "Hold this thought while I take care of this." Handle the interruption, briefly review the material you were just explaining so that students' minds are refreshed, then continue. If you establish this routine at the beginning of the year, you and your students will be able to manage interruptions with little distraction.

MONITORING LEARNING DURING INSTRUCTION

Students learn best when they are actively engaged in a lesson. In a perfect math class, the attention of your students is entirely directed toward what you are saying and doing, and they are enthusiastically participating in individual, group, or class exercises. Their minds are absorbing, analyzing, and organizing information, recognizing relationships and patterns, and drawing conclusions. They are learning math.

As any math teacher (or any other teacher) knows, keeping a classroom of students focused completely on a lesson for a whole period is daunting. No matter how well designed and interesting a lesson may be, some students will require close but subtle monitoring to make certain their attention remains on the topic throughout the class. You can help your students keep on task as your lesson progresses by periodically checking for learning.

Consider the following strategies:

- Throughout the instruction or activity, ask students specific questions that require them to summarize or review the concepts and skills you are presenting. Their answers will show you if they understand what you have taught so far, and reinforce the material for their classmates.

- Encourage your students to complete sample problems of new material on the board, demonstrating that they understand and can apply what they've learned.

- Ask questions that require your students to predict relationships or patterns. If they can, you know that they not only understand the material but they have extended their learning.

- When students are working in groups, circulate around the room and sit in on different groups for a few minutes. This will allow you to monitor the group's progress, determine if students understand the material, and allow you to help them work through any confusion.

- Periodically ask students if they have any questions. Answering their questions as you go along keeps them involved in the lesson.

- Ask students to state how the topic relates to other topics in math, other subjects, and their lives. This helps them to expand their learning by connecting math to other areas.

Based on the responses of your students, you will know if they are having trouble grasping the new material of the lesson. You can then adjust the lesson, provide needed background information, or reteach specific skills or concepts. By consistently monitoring your students' learning during the day's lessons and activities, you will help ensure that they are benefiting from your instruction—and you will avoid the disappointing experience of having delivered a complete lesson only to realize that your students did not meet the objectives you had set for them.

VIDEOTAPING YOUR DELIVERY AND BUILDING CONFIDENCE

An excellent way to assess your methods of instruction is to videotape yourself or ask a colleague to tape you while you are presenting a lesson. Afterward, as you view the tape, you can discover both the strengths and weaknesses in your delivery and classroom presence. By refining your strengths and improving any weaknesses, you will enhance your instructional skills and build confidence.

Note: Before you videotape yourself, check with your supervisor about school policy regarding videotaping. There may be restrictions, particularly regarding taping of students. If there are, perhaps you can arrange for the camera to record only you.

As you watch a video of your teaching techniques, ask yourself the following questions:

- Did I look professional?
- Did I speak clearly and with enthusiasm?
- Did I speak to the level of my students? Were my explanations logical and easy for my students to understand?
- Did I avoid fillers such as "like," "um," and "ah"?
- Did I use vocal emphasis to highlight important points?
- Did I use the correct math terms?
- Did my body language convey confidence? Did it match my words?
- Did I write clearly on the board, or project information clearly on the screen?
- Were all of my directions clear?
- Did I make eye contact with students?
- Did I teach to all parts of the room?
- Did I handle technology efficiently?
- Did I involve numerous students during my presentation by asking questions and encouraging their participation?
- Did I use a variety of teaching techniques and address different learning styles?
- Did I monitor learning as the lesson progressed?
- Was I aware of what students were doing as I presented the lesson?
- Did I interact positively with my students?
- Did I use time wisely?
- Did I manage transitions smoothly?

In the video, you will see yourself as you appear to your students. Careful viewing and analysis of the video will help you to improve the delivery of your instruction, which will enhance the effectiveness of your lessons.

Through honest self-evaluation and practice you can improve your skills for delivering instruction. By presenting engaging lessons you will be able to capture and maintain the interest of your students, which is a prerequisite for learning.

Quick Review for Providing Effective Math Instruction

Providing effective instruction day after day is hard work. Even though you may plan a great lesson, if you are unable to present it in a manner that your students understand, you are unlikely to meet the lesson's objectives. Potential learning will be undermined.

The following tips will help you to deliver instruction that your students will find motivating and interesting, thus creating a positive learning experience:

- Be willing to assume the many different roles of a math teacher.
- Use various methods of instruction that address the diverse learning styles of your students.
- Foster the development of problem-solving skills in your students.
- Motivate your students by presenting interesting lessons that are relevant to their lives.
- Improve your instructional delivery skills through clear speaking and positive body language.
- Handle math manipulatives effectively.
- Integrate technology with lessons and activities, and use the equipment with competence.
- Calmly and efficiently manage the common interruptions that disrupt lessons.
- Keep students on task during lessons and activities by encouraging their participation and consistently monitoring their learning.
- Consider videotaping your delivery of instruction, analyze your strengths and weaknesses, and work to improve your skills.

A well-designed math lesson will have little impact on learning if it is delivered through mediocre instruction. Likewise, effective instruction cannot make a mediocre lesson a successful learning experience for students. Both elements—well-designed lessons and effective instruction—are necessary for students to achieve the objectives of your lessons.

SECTION TWELVE

Evaluating the Progress of Your Students

Evaluation is the means through which you can determine if your students have met your objectives for a lesson, chapter, or unit. It is also the basis of your grades, which act as a summary of your students' progress for the students, their parents and guardians, administrators, and other officials, such as awards committee members and college admission officers. Thus, evaluation can have significant impact beyond the walls of your classroom.

You can evaluate your students in a variety of ways, including formal assessments, such as examinations, tests, and quizzes; homework; classroom observations of individual and group activities; math notebooks; portfolios; and conferencing with students. Although some assessments, such as homework and classwork, enable you to evaluate your students' daily progress, others, such as standardized tests and final examinations, enable you to evaluate your students' long-term progress. Evaluation is an essential and ongoing part of a productive math class.

To be a valid indicator of the progress of your students, evaluation must be accurate, fair, and consistent. It must focus on what has been taught, identifying the skills and concepts that have been mastered as well as the material on which students need more work, and should assess a variety of levels of knowledge. Effective evaluation is a measure of the success of your students, and a reflection of the success of your teaching.

Devising a Fair System of Grading

Your grading system provides the framework for the evaluation of your students. Although most school districts and math departments have grading policies, many also allow their teachers flexibility in grading. If you devise your own grading system, you need to carefully consider the activities you feel are most important in your math program. Perhaps you will decide that your evaluation of your students will be mostly based on tests and quizzes. Or maybe you feel that your students will be best evaluated through projects, classwork, and observation, with less emphasis on tests and quizzes. Prioritizing the different types of activities enables you to assign a percent value to each.

Following are examples of possible grading systems. You may select one or use them as guides to create your own:

Sample One	Sample Two	Sample Three
Tests, 30%	Tests, 20%	Tests, 25%
Quizzes, 20%	Projects, 20%	Projects, 25%
Projects, 10%	Portfolios, 20%	Quizzes, 15%
Math notebooks, 10%	Math notebooks, 10%	Math notebooks, 15%
Homework, 15%	Homework, 10%	Homework, 10%
Classwork, 15%	Classwork, 10%	Classwork, 10%
	Effort, 10%	

Whatever the form of your grading system, you must be cognizant of and make provisions for your special-needs students. Some of these students may have an IEP (Individualized Education Program) or a 504 plan, which you are bound by law to follow. Such plans may have stipulations about testing or implementation of a specific activity. For example, a student's IEP may stipulate that he always be allowed to use a calculator in class, or that he be permitted more time for testing. You must be familiar with any IEPs and 504 plans for your students and implement their requirements. You may also need to modify your grading system for students who speak little or no English. To hold these students accountable for all of the activities of your class may be unrealistic, particularly for material that requires reading. A grading system must be fair to all the students in a class.

Your grading system should include various methods of assessment. It should correlate with your objectives and enable you to determine the overall progress of each student in your class.

Ways to Assess Student Learning

Assessment can be formative or summative. Formative assessment is part of the instructional process. It gives you information about student understanding as they are learning and enables you to make adjustments in instruction. Observation of student learning during a classroom activity is an example of formative assessment. Summative assessment, such as a unit test, identifies what students know and do not know at a specific point. It summarizes student learning in relation to content. Both formative and summative assessments should be a part of your program.

There are many ways you can assess the learning of your students. Using a variety of assessment tools enables you to consider a range of activities and learning preferences, providing you with sufficient material to evaluate your students fairly.

ASSESSMENT THROUGH TESTS AND QUIZZES

For most math teachers, tests and quizzes are the staples of assessment. Whereas well-designed tests and quizzes can be excellent tools for measuring student learning—identifying mastery of specific objectives, providing insight to students' mathematical reasoning, and revealing the depth of students' understanding of math concepts—mediocre ones usually provide little information other than a grade.

To create effective tests and quizzes, consider that a valid math test or quiz satisfies the following:

- Has clear directions
- Includes a point distribution, such as "Part 1 (10 points)," which informs students how much each part of a test or quiz is worth and enables them to estimate their progress as they work
- Accurately measures the objectives it was designed to measure
- Has accurate and clearly labeled figures, diagrams, tables, and charts
- Focuses on material and problems students have worked on in class and completed for homework
- Contains various types of questions, for example, computation, word problems, multiple choice, fill-in, and open-ended problems that require students to explain their reasoning
- Provides questions that address various learning styles
- Includes questions that assess several levels of knowledge, including critical-thinking skills
- Contains enough questions to provide you with an accurate measure of student progress, but not so many questions that students do not have enough time to finish during class

In addition, consider these points:

- Help your students prepare for tests by providing reviews and study guides that highlight important topics, concepts, and skills.
- Hand out different versions of the same test or quiz randomly to students to reduce opportunities for cheating.

- Be sure to save copies of tests and quizzes to adapt for use in following years. Saving them electronically makes updating them easy.

- If a student is absent on the day of a test or quiz, give him a different version when he returns.

- Always grade tests and quizzes promptly. Returning them the next day (provided all students have taken the test) gives students feedback while the material is still fresh in their minds. You may instruct students to correct the problems that they got wrong. If necessary, go over the problems that gave them the most trouble.

Tests and quizzes provide valuable information for you and your students. They can help you assess student knowledge, and enable you to decide if it is necessary to reteach vital skills and concepts before moving on to the next unit. They also offer insight into your students' problem-solving skills. Finally, tests and quizzes let your students know "how they are doing," and what they still need to learn.

ASSESSMENT THROUGH OPEN-ENDED PROBLEMS

An open-ended problem has one or more questions that require students to think critically in order to find a solution. If there is more than one question to the problem, the questions usually advance in degree of difficulty from simple to challenging. The answers to open-ended problems often require explanations or justification of the students' answers and responses. Sometimes students must construct or interpret a table, diagram, figure, or chart to support the validity of their conclusions. Open-ended problems provide teachers with insight and understanding of their students' mathematical reasoning, rather than merely knowing that students got the right answer.

Rubrics are useful for scoring open-ended problems. (A rubric is a method of grading according to a given scale. Rubrics help to ensure that all student responses are scored in the same manner, using the same criteria.) Giving your students the rubric you will use for scoring informs them how their responses will be graded, which in turn guides them in formulating their answers. You might even provide sample problems and responses and allow your students to grade them. This not only familiarizes your students with the scoring methods but enables them to analyze problem-solving strategies and communication of mathematical ideas.

Following is a sample rubric for scoring open-ended math problems:

3 Points: The response shows complete understanding of the math concepts. The response is detailed, clear, and contains few, if any, errors. Valid reasons for the mathematical procedures used are provided.

2 Points: The response shows near complete understanding of the math concepts. Most of the mathematical procedures used are correct; however, there are a few minor errors. Reasons for the mathematical procedures used require more explanation.

1 Point: This response shows limited understanding of the math concepts. The explanation contains major errors and the procedures used are incomplete or inaccurate.

0 Point: There is no evidence of understanding of the math concepts. The response contains major errors and the procedures used are incorrect. There may be no explanation or the explanation makes little sense.

You may use this rubric as presented or use it to design a rubric of your own. Note that many state departments of education use specific rubrics for scoring open-ended problems on their state math assessments. If your state is among these, use the same rubric for assessing the open-ended problems you assign your students. As they become accustomed to the rubric, they will know what is expected of them and how their work will be assessed. You may also create rubrics of your own, listing the specific information that students must provide for earning a specific score.

To find examples of rubrics, check the Internet with the search term "math rubrics." Because they help to ensure consistency and fairness, rubrics are useful scoring tools for open-ended problems.

ASSESSMENT THROUGH GROUP ACTIVITIES

Working in groups offers many advantages to students. While collaborating on investigations of math concepts, solving complex problems, or exploring mathematical properties through manipulatives, students are able to share ideas with the support of other group members. For many students, working with others enables them to achieve better results than working alone.

Group activities present you with grading options. You may choose to assign students a group grade, an individual grade, or a combination of both. To help you credit the contributions of students fairly, you may invite them to share in the evaluative process by asking them to state what each member contributed to the group's efforts. Having each group member write an explanation of his specific contribution to the group's work and what he learned from the activity will also help you assess each student's contributions, as well as the quality of contributions.

To assess group work with consistency and fairness, you should create a rubric that concentrates on the aspects of the activity that support your objectives. Following is a list of criteria that you might use in designing a rubric for assessing a group activity:

◎ Students identify the problem.
◎ Students use appropriate mathematical language during the activity. They use precise rather than general terms.

- Students brainstorm to generate possible strategies for finding a solution.
- Group members experiment with different strategies in seeking a solution.
- Various conjectures are tested.
- The best strategy for finding a solution is selected.
- The group remains on task and on topic.
- Students use technology where appropriate.
- Students gather the necessary data.
- Students ignore unnecessary information and facts.
- Students analyze and organize data.
- Models are used as necessary.
- Reasonable solutions are found.
- Solutions are justified.
- All members of the group contribute in a positive manner.
- Findings are presented clearly.

These, of course, are just some criteria you might consider for creating a rubric to evaluate a group activity. After selecting specific criteria based on the activity, designate a point value for each one. Sharing the rubric with your students before they start working informs them how they will be evaluated and what they should try to achieve. (See "Group Work" and "How to Work in a Math Group" in Section Seven.)

ASSESSMENT THROUGH MATH NOTEBOOKS

Math notebooks can be useful tools for students, serving as a resource containing notes, classwork, homework, tests, and quizzes. Many math teachers consider notebooks to be so important that they regularly review and assess them. When your students know that you will periodically check their notebooks, they are more likely to keep them up to date, thus ensuring that the notebooks remain a useful study aid.

If you decide to assess math notebooks, you should consider creating a rubric consisting of the criteria you feel best support your purpose for requiring students to maintain notebooks.

Following is a sample rubric, based on a total of 10 points:

- Agenda or assignment pad (assignments are dated in order, correct, and current): 1 point
- General organization (set up according to your instructions): 1 point
- Class notes (in order by date and completed): 2 points

- Classwork assignments (in order by date; appropriately labeled; completed; incorrect answers are corrected): 2 points

- Homework assignments (in order by date; appropriately labeled; completed; incorrect answers are corrected): 2 points

- Quizzes and tests (in order by date; incorrect answers are corrected): 2 points

A math notebook indicates the degree of a student's involvement in your class. It shows her commitment to studying math and is therefore a valid tool for assessment. (See "Math Notebooks" and "Tips for Keeping a Math Notebook" in Section Seven.)

ASSESSMENT THROUGH MATH PROJECTS

Math projects serve as an excellent alternative to traditional assessments. Projects allow students to solve real-life problems and demonstrate skills that go beyond those measured on a typical test.

To ensure consistent and effective evaluation of math projects, you should create a rubric based upon your objectives for your students. While you can designate point totals to fit your overall grading system, the sample rubric that follows totals 100 points, which can be easily translated to percentages and grades. (The rubric is based on the sample rubric in *Hands-on Math Projects with Real-Life Applications, Grades 6–12*, Second Edition, by Judith and Gary Muschla, Jossey-Bass, 2006, page 38.)

Satisfactory Solution: The solution is valid and practical. 25 points

Justification of Results: Students justified results through an oral presentation, written report, or discussion. They backed up the results with sound mathematical arguments. 15 points

Methods: Students eliminated impractical procedures and focused their efforts on the most useful. If necessary, students eliminated irrelevant data and found needed data. They were able to analyze and organize information and use technology where applicable. 15 points

Accuracy: Reasoning and computation were logical and accurate. 15 points

Creativity: Students showed original or insightful thinking. 10 points

Persistence: Students did not give up. 10 points

Cooperation with Group: Students worked well together, shared ideas, and listened to each other's ideas. They showed a willingness to help each other. 10 points

Math projects are long-term activities that require students to demonstrate a variety of skills. A rubric that considers the different skills and behaviors necessary for the satisfactory completion of a project is a fair means of assessment.

ASSESSMENT THROUGH WRITING

In recent years, writing has assumed greater importance in math curriculums. Most math teachers evaluate many types of student writing, including formal research reports, explanations of solutions to open-ended problems, essays, and journal entries. (Note: We do not recommend grading journal entries in which students share their thoughts and feelings about math. If these entries are graded, many students will write what they think will result in a higher grade rather than their honest reflections about an issue, topic, or problem in math. Students lose the chance for free-spirited thinking about math and teachers lose the chance for gaining unfiltered insight to their students' impressions about math.)

When assessing the writing of your students, select specific criteria to guide your evaluative efforts. Along with providing a score or grade, you should offer comments and suggestions. All writers enjoy responses to their work.

Following is a sample rubric you can use for evaluating the writing of your students. (This rubric is based on the sample rubric in *Hands-on Math Projects with Real-Life Applications, Grades 6–12,* page 39.)

- *Focus*: The topic is clearly defined. All ideas support the topic. 10 points
- *Content*: The student uses fresh, insightful, or original ideas. The topic is developed and supported with details. Mathematical reasoning is sound and shows an understanding of concepts. 25 points
- *Organization*: The piece progresses logically from beginning to end. For long pieces, an introduction, body, and conclusion can be identified. 25 points
- *Style*: The writing is appropriate for the topic and audience. Ideas are communicated effectively. There is a distinct voice. 15 points
- *Mechanics*: The writer uses correct punctuation, grammar, and spelling. 25 points

You might consider consulting with your students' English teacher regarding the development of their writing skills. She might be willing to share guidelines for the way she teaches writing. Working together, you and she can provide consistent writing instruction to students.

ASSESSMENT THROUGH MATH PORTFOLIOS

A math portfolio is a collection of examples of a student's work. A portfolio may include a variety of items, including: do-nows, journal entries, assignments, quizzes, tests, solutions to open-ended problems, reports, projects, and research papers. Portfolios can highlight a student's overall progress in math, as well as his or her attitudes about math.

Although some teachers prefer that students maintain a work portfolio, which serves simply as a storehouse of work and is not graded, portfolios may also be used for assessment. Such assessment portfolios are evaluated according to specific criteria, which are based on the teacher's instructional program and objectives. Encouraging your students to select what they feel is their best work to include in an assessment portfolio gives them input in the process along with the opportunity to review and evaluate their work.

Following are examples of criteria that you might use to assess math portfolios. The work in a portfolio should demonstrate that the student:

- Understands specific math concepts and skills
- Uses the appropriate math terms
- Applies a variety of problem-solving strategies
- Makes and tests conjectures
- Verifies results
- Draws and supports conclusions
- Selects the appropriate technology for solving problems
- Clearly communicates ideas about mathematics
- Uses critical thinking

Selecting criteria that address specific objectives will help to ensure your objectivity when assessing the portfolios of your students. Portfolios provide a long-term view of your students' growth in math and can be an important means of assessment.

ASSESSMENT THROUGH CLASSWORK

Most teachers consider classwork to be all of the work students complete in the classroom. Although the value that teachers assign to classwork in their overall grading systems may vary, classwork is an effective and relatively simple way to evaluate daily learning. For example, you may evaluate the work students do in class by monitoring their activity, or you may collect their work upon completion and check it later.

As most school districts and math departments have policies for assessing classwork, you should base your assessment method on administrative guidelines. In the absence of guidelines, you must determine how you will assess classwork and how you will average classwork scores with the scores of your students' other work.

Following are two practical methods:

- Assign 3 points for excellent work, 2 points for average work, 1 point for below-average work, and 0 if a student does not hand in the assignment. The total points a student earns over the course of a marking period would then be included with his scores for other work.

○ Mark acceptable work with a check, unacceptable work with a check minus, and missing work with a zero. Determine a point value for checks and check minuses and include the totals with your other scores.

By regularly checking the classwork of your students, you remain aware of their daily progress and are able to provide immediate feedback. If you find that students are having trouble with specific concepts or skills, you may reteach the material, helping students to gain mastery before moving on to other topics. (See "Classwork" in Section Seven.)

ASSESSMENT THROUGH HOMEWORK

Homework provides yet another means for assessment. Similar to classwork, the value that teachers assign to homework varies. Some teachers do not grade or even collect homework, believing that homework is practice for students. Moreover, as homework may be copied, or, in some cases, done by well-meaning parents, its value as an assessment tool is limited. Others, however, assign a point value to homework that they average with their students' overall grades.

Before deciding on a value for homework, familiarize yourself with your school district's or math department's policies regarding the assessment of homework. Base your assessment methods on the appropriate guidelines. Whether you choose to collect and correct homework each day (this can quickly become overwhelming), collect and spot-check homework, or circulate around your room at the beginning of class and quickly note which students completed their homework, your effort at checking homework informs students that you consider the completion of homework to be an important part of your course.

As with classwork, you may use a simple point system for assessment: 3 points for excellent work, 2 points for average work, 1 point for below-average work, and 0 for missing homework. Instead of points, you may use a check for acceptable homework, a check minus for unacceptable work, and a zero for missing work. Determine point values, and at the end of the marking period include scores for homework with your students' other scores.

Regularly checking homework enables you to closely follow the progress of your students and reteach material that proves to be troublesome to them. When you assess homework, you are able to provide students with immediate feedback as to how well they are doing. (See "Homework" in Section Seven.)

Preparing Your Students for Standardized Math Tests

The administration of standardized tests is an annual event in schools, with the test results having major implications for individual students, schools, and school districts. Scores indicate whether students are meeting state standards, they allow for a comparison of district students to students across the nation, they may be used to determine a student's placement in advanced or remedial classes, or, in the case of high school seniors, they may be used as a criterion for college acceptance. Whether one agrees or disagrees with the purpose and validity of standardized tests, the public as a whole looks upon the scores as a measure of how well students and their schools are doing. Because of the significance of test scores, students should be prepared for testing.

The first step toward preparation is for you to check with your math supervisor or administrator in charge of testing. He or she may have support materials for the test your school will be giving. Many publishers of standardized tests have sample materials and practice tests that are designed to give students an idea of what to expect on the actual test. Such materials can help students anticipate the kinds of problems they will encounter, familiarize them with the test's format, and relieve unnecessary anxiety.

Because most school districts administer standardized tests that support state standards and objectives in specific subject areas, you should use your state department of education as a resource in your preparation efforts. Many states provide guidelines and sample problems to aid students in preparing for tests.

Following are strategies for ensuring that your students will excel on standardized math tests:

- ◎ Develop a math program that is based on your state standards. When your instruction supports state objectives, which in turn are at the heart of the standardized test your students will take, you will be preparing students for the test throughout the year.

- ◎ Assign sample problems to prepare your students for the test.

- ◎ Provide your students with any rubric that will be used to score open-ended problems. Such rubrics, if available, can be obtained from the Web site of your state department of education. If you are not using a statewide test, you may be able to obtain a rubric from the preparation materials for a specific test.

- ◎ Give your students practice in using the test's format. As most standardized tests use answer sheets, provide some tests of your own in which students must darken circles on an answer sheet.

- Provide any reference sheets that students may use during the test. Encourage students to be familiar with the contents of reference sheets, such as formulas, equivalencies, and properties.

- If your students are permitted to use calculators during the test, make certain they have ample opportunity in class to learn how to use the same type of calculator they will be using on test day.

- Throughout the year, give some tests with time restraints. This will give students practice in pacing and budgeting their time.

- Practice reading directions on tests and test items to familiarize your students with test terminology and to help them avoid misinterpreting instructions on standardized tests.

- Provide your students with information about the scoring of multiple-choice questions. Let them know if there is a penalty for guessing. If there is, students should answer a question only if they are reasonably sure they will be correct. If there is no penalty for guessing, students should answer all questions.

- Encourage your students to double-check their work on all class tests and quizzes. By promoting this habit, you will increase the chances that they will check their work on standardized tests.

- Speak positively of standardized tests. Encourage your students not to fear them, but rather look upon them as a chance to demonstrate their skills and ability in math.

To further prepare your students for the test, distribute and discuss the following "Tips for Taking Standardized Math Tests." Suggest that students retain this sheet in their binders and review it the night before the test.

Tips for Taking Standardized Math Tests

Name _____ Date _____ Period _____

You can improve your scores on a standardized math test by doing the following:

1. On the night before the test, get a good night's sleep. Go to bed a few minutes early and wake up with plenty of time to get to school.

2. Eat a solid breakfast on the morning of the test. If your test is in the afternoon, be sure to eat a good lunch.

3. Know where you are scheduled to take the test and what time the test starts. Be sure to arrive on time. Rushing to the test may upset you and interfere with your ability to concentrate.

4. Be sure you have all of the materials you will need: at least two sharpened No. 2 pencils, erasers, calculator with extra batteries, glasses or contact lenses, and tissues if you have allergies or a cold.

5. Listen carefully to all directions. If you do not understand something, ask your teacher or the person administering the test.

6. Write your name and other identifying information on your answer sheet correctly.

7. Read all questions and their possible answers. Reread them if necessary. Concentrate. Sometimes an answer that at first seems right is in fact wrong.

8. Pace yourself. Be aware of any time limits.

9. Try not to make careless mistakes. Stay focused during the test.

10. Make certain you are putting your answers in their correct place on the answer sheet. (If it is permitted, use an index card or a piece of scrap paper to keep your place.) Make certain you fill in answers darkly and neatly. If you change an answer, erase it entirely.

11. Try to answer the questions in order, but do not spend too much time on any one question. Skip hard questions that give you trouble and go back to them if you have time later. Remember, for every hard question on a test, there are many easy ones.

12. Answer all parts of open-ended problems completely with detailed explanations.

13. If you cannot decide between two or more answers for a question, compare the answers. Eliminate those that seem less likely to be correct. If there is no penalty for guessing, be sure to answer all of the questions. If there is a penalty for guessing, only answer questions that you are reasonably sure you will get right.

14. Double-check your work if you have time. Make sure each answer is reasonable and makes sense.

15. Be confident and think positively. Be ready to do your best. Students who are confident, try hard, and believe they will do well usually do better than those who are worried about the test.

When you prepare your students for a standardized test, you are not merely giving them sample problems and offering test-taking advice, you are demonstrating to them the importance of the test and building their confidence. They are more likely to come to the test with a positive attitude and ready to try their best than students who have had no preparation. They are also more likely to attain higher scores than they would had they been unprepared.

Evaluating Assessment Results

All kinds of tests provide useful information about your students, your course, and your instruction. Whether for standardized tests or the tests you give your students, evaluating assessment results can help you improve your teaching and math program.

One of the benefits of standardized tests is the assessment data that is reported with the scores. The data typically provides individual student profiles that display detailed test results for each student. It also offers comparisons of the scores of your students to the scores of students of similar communities as well as national norms. In addition, most tests provide a percentile rank, which compares a student's score with the scores of other students who took the same test. A percentile rank of 90, for example, means that the student scored better than 90 percent of the other students who took the test. A percentile rank of 20 means that he scored better than only 20 percent of the other test takers. The assessment data that accompanies the scores of a standardized test can help you understand a student's overall capability in math.

Evaluating assessment results of standardized tests also enables you to find the strengths and weaknesses of your math program. Standardized test results are often broken down according to math standards. By examining these results, you can determine which areas in your program require improvement. You can then review your curriculum and teaching methods. Perhaps you need to concentrate more instruction on particular topics or objectives. Maybe your curriculum needs to be revised to more closely align with state standards. Or possibly you need to differentiate instruction more, modifying lessons to better satisfy diverse learning styles. Honest evaluation of assessment data can help you to improve your instruction.

The results of the classroom tests you give your students can also give you valuable information. If several students fail the same test, or if many students have difficulty with the same part of a test or with specific questions or problems, you need to ascertain the reasons.

Asking yourself questions such as the following can help you find answers:

- Was there a problem with my instruction?
  - Did I adequately cover the material in a way that addressed the needs of all learners?
  - Did I provide enough time for instruction?

- Did I provide enough practice?
- Did students have the prerequisite skills needed to master the material?
- Did I relate the material to the students' lives, making it easier for them to understand?
- Could I have presented this information differently? If yes, how?
- Did I provide review or study guides that helped students prepare for the questions they had to answer on the test?

◎ Was there a problem with the test itself?

- Were the problems on the test similar to those covered in class and for homework?
- Did the test accurately reflect the objectives of the unit?
- Were the directions and questions worded clearly?
- Were students familiar with the format of the test?
- Did I provide enough space between problems for students to view them easily?
- Was the test too long? Were students able to finish? Did they rush to complete it because they did not have enough time?

◎ Was the problem with the students?

- Was the test given before a vacation or other event that could have distracted students?
- Did students study for the test?

Once you have identified possible reasons for the unsatisfactory results, address them. Use assessment results to improve your instructional program and help your students achieve success in your math class.

Record Keeping

Not too long ago, all teachers used paper grade books for recording the grades of their students. Today, most teachers use electronic grade books. Many record their grades electronically but also maintain hard copies in paper grade books.

Following are tips for maintaining your grades either in a traditional or electronic grade book:

◎ Follow the specific policies of your school for maintaining grades. Your grade book is a legal document, and you must maintain it accurately.

◎ Set up the dates and students' names for a new marking period prior to the start of the marking period.

◎ Record attendance for each class.

- Include grades from a variety of sources—classwork, homework, tests, quizzes, math notebooks, and projects—to provide a fair and accurate assessment of student progress.

- Inform students and parents or guardians how you will determine students' grades. This should be done at the beginning of the year. (See "Devising a Fair System of Grading" earlier in this section.)

- Label all assignments so that you can easily recognize them, for example, "Unit Test 3" or "HW, p. 106, 1–10" (Homework, page 106, problem numbers 1–10). Be sure to provide a key if you use any shorthand.

- Keep the grades in your grade book current. Grade papers as you receive them and enter them promptly. Avoid letting papers pile up.

- Make sure that you have an effective distribution of categories. For example, do not give only one quiz if quizzes count 20 percent in your grading system.

- Remember that grades are confidential. Never share a student's grades with another student or a parent or guardian other than the student's own.

Electronic grade books have various capabilities. Along with recording and averaging grades, many programs allow you to create reports of missed work, generate graphs of student performance, make seating charts, and allow parents and guardians to view their child's grades online. Most school districts purchase the grading program they require their teachers to use. However, if your school permits teachers to select their own electronic grade books, you might consider MyGradeBook (www.mygradebook.com) or GradeBookWizard (www.gradebookwizard.com), both of which enable teachers to manage grade books online. For more information and other electronic grade books, search the Internet with the term "electronic grade books."

The following tips apply to electronic grade books:

- Electronic grade books are computer programs. As with any program, you must be willing to learn how to use it. Attend any in-services or workshops your school offers to familiarize teachers with the program. Utilize any tutorials that come with the program, and be sure to keep your manual handy and refer to it as necessary.

- When you first begin using an electronic grade book, use only its basic functions. Avoid trying to do too much too soon. As you become more familiar with the program, you can use more components.

- Set up your category weights to reflect your grading system.

- If your grade book has a comment section, be sure that your comments are professional and appropriate.

○ Keep your grade book up to date. This is particularly necessary if parents and guardians can view their child's grade online. Set aside a regular period of time to enter grades.

○ Keep your grade book secure. Use passwords to prevent unauthorized access.

○ Always back up your data electronically and with hard copies.

○ Ask for help if you have a problem. The technology person in your school or another teacher may be able to answer your question easily, saving you the time you would spend trying to find the answer yourself.

If your school permits teachers to include comments on report cards, remember that comments can be easily misinterpreted. Avoid overly blunt or harsh statements. Focus your comments on how students can improve their math grades, and always try to include a positive comment for every student. Following are some examples of appropriate comments:

○ . . . always comes to class prepared with the necessary materials.

○ . . . contributes to class activities and discussions.

○ . . . needs to prepare for tests in order to realize her full potential in class.

○ . . . is capable of greater achievement; he needs to complete and hand in assignments consistently.

○ . . . notebook is complete and up-to-date.

Your grade book, whether it is paper or electronic, is the record of your students' progress and the source of their report card grades. Because report cards are a means of communicating with parents and guardians, the grades must always accurately reflect the student's achievement in your class.

IF YOUR GRADES ARE CHALLENGED

Because of their importance—to honor rolls, class rank, and potential awards—there will be occasions when students, parents or guardians, or even administrators will question the grades you have given. You can reduce the frequency of challenges to your grades by following your district's grading policy, informing students, parents, and guardians of your grading system at the beginning of the year, providing various types of assessment, and informing parents and guardians promptly if their son or daughter is experiencing difficulty in your class. When one of your grades is challenged, your response should be both considerate and professional. It certainly should include valid reasons for giving the grade you did.

Here are some tips:

- Do not react negatively to the challenge, or view it as a personal attack on you, the class, or your teaching.
- Focus on the reason for the challenge. Listen carefully to determine why the grade is being challenged. Following are two examples:
  - If a student challenges a grade on a test, ask which problem marked incorrect he believes is right. Ask why he believes his answer is right. Go through the problem with him and show him his mistake.
  - If a student claims that her average should be higher, implying that you entered scores incorrectly in your grade book, ask her to show you the grades. (This, of course, is a major reason why students should maintain math notebooks and keep graded papers.)
- When grading papers, for example open-ended problems, write comments that indicate why you graded the paper as you did.
- If you find that you made a mistake in grading, graciously correct the error. This will show that you are human, receptive, and able to admit an error. After all, everybody makes mistakes.

Maintaining your grade book is vital to ensuring that you accurately record the progress of your students. When you keep your grade book current, you are able to monitor your students' progress and respond quickly to those who might need additional help.

Quick Review for Evaluating the Progress of Your Students

Evaluation of your students' work enables you to determine if they have mastered the objectives of your math program. Because evaluation is a clear indicator of a student's learning and the basis of report card grades, it is among your most serious responsibilities as a teacher.

The following points highlight steps you should take to evaluate the progress of your students:

- Implement a fair and consistent system of grading, based on the objectives of your math program.
- Assess various kinds of student work—for example, classwork, homework, tests, quizzes, open-ended problems, group activities, math notebooks, projects, and portfolios.
- Prepare your students for standardized tests by developing a math program based on state standards, providing sample problems, and giving them practice using the test's format.

○ Use the results of assessments to identify material your students found difficult to master, reteach weak skills, and improve your instructional delivery skills.

○ Maintain an accurate and up-to-date grade book. If you use an electronic grade book, maintain backup copies as well as hard copies of students' grades.

○ Reduce the potential for challenges to the grades you give your students by following your district's grading policy, providing a variety of assessments, making sure that students, parents, and guardians understand your grading system, and keeping parents and guardians updated on the progress of their children.

Students often look upon evaluation—particularly their grades—as simply an indicator of how they are doing in your class. Evaluation, however, is much more. Through careful evaluation of your students' progress, you can determine areas in which they are excelling, and also those areas where they need more work. Evaluation can be a useful tool in tailoring your instruction to meet the needs of all the students in your class.

SECTION THIRTEEN

Managing Inappropriate Behavior

*E*very teacher, every day, must manage inappropriate behavior on the part of some students. Effective teachers maintain control of their classrooms and do not allow the behavior of these few to undermine the learning environment for the many. Such teachers are in charge, and their students respect their authority.

It is not easy to control a classroom of students, many of whom have concerns other than math. Perhaps because of a lack of parental supervision, a student regularly comes to school late and misses part of your first-period class; because she feels neglected at home, a student misbehaves in class in an attempt to gain your attention; or because of living in a neighborhood where violence is prevalent, a student is aggressive and prone to fighting. These and similar behaviors can make classroom control a daily concern. Little learning can occur if classroom control is lacking.

The key to maintaining control of your classroom is to prevent as much inappropriate behavior as possible, and to address swiftly and effectively any negative behavior that occurs. You can significantly reduce behavior problems through consistent application of school and classroom rules, proactive intervention, and clear expectations, motivation, and encouragement. Through calm, steady, and practical actions, you can efficiently handle most problems that arise. When you maintain control of your classroom, you are creating the structure on which a productive and enjoyable math class can be built.

Addressing Inappropriate Behavior in Your Math Class

You have clear responsibilities to address inappropriate behavior. Your school's faculty handbook and discipline policies will include details of the scope of these responsibilities.

Most important, you must ensure a safe and orderly environment for your students. You must create practical rules and procedures, and inform your students of your behavioral requirements and of the consequences for breaking the rules. You must monitor your class to make certain that students are behaving properly; never leave the class unsupervised. When students act inappropriately, you must correct them promptly. Consequences should be fair and consistent, and you must treat all students equally.

Most misbehavior can and should be handled in your classroom. When students see you managing problems, you gain their respect. They will look upon you as the authority figure in the class. If, however, you pass issues off onto someone else, students will conclude that you either are unable to take action yourself, or do not care to do so. They may respect the vice principal you send them to, but they may not respect you, and this will only make controlling the classroom more challenging for you.

If your school has a handbook with policies for dealing with disciplinary issues, follow its guidelines closely and use the guidelines as the foundation of your classroom rules. In responding to inappropriate behavior, you should base your actions on the seriousness of the behavior and its effect on learning, both for the offending student and others.

Minor misbehavior, such as when a student is not paying attention to your lesson but is not bothering anyone else, can be handled with a simple, "Brian, I need your attention here." For other minor problems, you may choose to step close to a student's desk, give him the "teacher's look," or ask him to stop what he is doing. For most instances of minor misbehavior, such simple actions are all that you will need to do.

For more serious problems, such as when a student constantly fails to complete her work, you may need to have the student report to you after school to discuss the problem. Explain to the student why you asked her to report after school, discuss the importance of completing work, and encourage her to act positively. You might also discuss further consequences, such as notifying her parents or guardians if her behavior continues.

For very serious problems, or repeated problems, you will need to follow through with consequences detailed in your school's discipline policies or your classroom's rules. For example, to discourage repeated tardiness you may assign a day's detention, to end a student's severe teasing of others you may need to contact parents or guardians, or to control a student's continuous disruption you may need to refer the matter to the appropriate administrator. (See "Involving Parents and Guardians in Addressing Inappropriate Behavior" and "Involving Administrators in Addressing Inappropriate Behavior" later in this section.)

Whichever way you respond to inappropriate behavior, you must act promptly and decisively. When you ignore problems or respond to them in an indecisive manner, students will assume that you are not sure of yourself and may begin to ignore your classroom rules.

Following are suggestions for addressing inappropriate behavior in your class:

- Learn about your students—their likes and dislikes, their strengths and weaknesses. When you understand your students, you are able to recognize potential problems and take steps to prevent misbehavior. Knowing your students also enables you to react quickly to problems and keep minor issues from escalating.

- Treat all students equally. Even students who are disciplined the most recognize and respect fairness.

- Establish reasonable and consistent classroom rules and consequences. Make sure that your students understand the rules; remind them of the rules as often as necessary. Displaying rules on the bulletin board is helpful.

- Be proactive. Try to avoid situations that might cause a problem. Seating two students next to each other even though they share intense mutual dislike in the hope that they can learn to get along can be a mistake.

- Speak to students calmly. Discuss what they did and explain why their behavior is unacceptable. Talk about how they could have behaved differently.

- Remember that you are responsible for the safety of the students in your class. In the case of a student threatening another student or becoming violent, you must act quickly. Follow the policies of your school.

For suggestions for managing specific behaviors, see "Common Examples of Inappropriate Behavior and How to Handle Them" later in this section.

You should always act in a positive manner when addressing inappropriate behaviors and avoid negative actions. Do not do the following:

- Ignore inappropriate behavior.

- Lose your temper or your composure. Some students will attempt to irritate you if they feel their actions can affect you.

- Negotiate with students. When you tell a student who is out of his seat for no valid reason to sit down, there should be no discussion. Permitting students to negotiate with you undermines your authority.

- Allow students to argue with you. That, too, will only undermine your authority.

- Embarrass students by strongly reprimanding them in front of other students. Instead, speak to the student quietly at her desk or in the hall. (Remember that you are still legally responsible for the students remaining in class and you should not leave the class unattended.)

- Shout at students or raise your voice excessively. With some students, shouting will lead to an argument.

- Threaten students if they do not behave. Threats, especially those you do not carry out, quickly erode your authority.
- Speak sarcastically or ridicule students in hopes of managing their behavior.
- Punish the entire class for the misbehavior of a few. Students will view this as unfair.
- Assign extra work as punishment. Students will resent the work and possibly come to dislike math.
- Allow yourself to dislike some students or treat them differently because of their lapses in behavior.

By the time they reach middle school (and certainly by high school!), students know how they should behave in class. Knowing how they should behave, however, does not always result in behaving as they should. When students misbehave, you must address the behavior promptly and effectively to minimize any disruption of learning.

Involving Parents and Guardians in Addressing Inappropriate Behavior

You will be able to manage most of the behavior problems you encounter simply by speaking with students and encouraging positive behavior. Despite your efforts, however, some students will find it difficult to comply with school and classroom rules. For these students, you will need to contact parents or guardians and request their support in solving the problem.

You may contact a parent or guardian by phone or e-mail. Following are some guidelines:

- If you contact a parent or guardian by phone:
  - Make sure you address or ask for the parent or guardian by the correct last name. (This avoids, for example, the awkward circumstance of greeting the former Mrs. Wilson as Mrs. Wilson after she has remarried and is now Mrs. Wallace.)
  - Identify yourself and mention the student's name. This is especially important if the student has siblings.
  - Explain that you are the student's math teacher and state the reason you are calling.
  - Explain the problem, what steps you have taken to solve it, and what the outcomes have been.
  - Ask the parent or guardian if she has any information that might help solve the problem. Sometimes a parent or guardian can share background or insight that can be useful.

- Request that the parent or guardian speak to the student about correcting the problem.

- Provide the parent or guardian with your phone number at school and a time you can be reached in case she wishes to contact you. (You may also provide her with your school e-mail address.)

- End the conversation by saying that you will call again if the problem continues.

- If you call and reach an answering machine, identify yourself, leave your name, and mention the student's name. Simply say that you are concerned about the student's recent behavior in school and that you would like to speak to the parent or guardian about it. Do not leave specifics on the message. Include your phone number at school and a time you can be reached.

- Keep a record of the discussion in case you need to take additional action. Note the date, time, to whom you spoke, and the action that you and the parent or guardian agreed to take. Later you can include the outcome.

- If you promised to follow up on the conversation, perhaps by calling again in two weeks, be sure to do so.

○ If you contact the parent or guardian by e-mail:

- Identify yourself and include the student's name. Explain the reason for the e-mail, briefly describe the problem and what you did to correct it, and what you would like the parent or guardian to do.

- Provide your phone number at school and a time he can reach you if he would like to discuss the problem in greater detail.

- Always be careful to say precisely what you wish to say and not say anything that can be misconstrued. Remember that e-mail serves as a written record.

- Print a copy of e-mail correspondences to keep as a reference.

For recurring or persistent problems with students that require you to contact parents or guardians repeatedly, you should maintain records of your conversations and meetings. Accurate records not only enable you to refer to past situations and actions, but provide details of problems should intervention by administrators, guidance counselors, or members of the child study team become necessary. You may find the form "Record of Parent-Guardian Contact" that follows helpful.

When you speak to parents or guardians, keep the conversation focused on their children. Although inappropriate behaviors often involve other students, and parents and guardians will often ask about what actions you are taking with these children, only discuss others in relation to the current issue. Do not extend the discussion beyond the specifics of the problem.

Record of Parent-Guardian Contact

Student's Name_____ Period_____

Parent's or Guardian's Name_____

Home Phone Number _____ Cell Number _____

E-Mail Address_____

Date	Type of Contact	Reason for Contact	Outcome	Follow-Up

THE VALUE OF BEHAVIOR CONTRACTS

Behavior contracts can be effective tools in encouraging appropriate behavior in students. These contracts can help clarify a problem and detail the steps that the student must take to improve his behavior. They send a powerful message that the behavior in question cannot continue.

Before implementing a behavior contract with a student, you should contact his or her parents or guardians, make them aware of the problem, and enlist their help in trying to solve it. The behavior contract should be kept as an option if other strategies fail.

Following are steps to take when considering using a behavior contract with a student:

- Meet with the student and his parents or guardians. Other teachers whom the contract might affect, such as an inclusion teacher, should also be present. Depending on the student and the problem, an administrator, guidance counselor, or member of the child study team might attend the meeting as well.

- Discuss the specific behavior that needs to be addressed. An example is a situation where the student does not complete assignments on time. Make certain that everyone understands the purpose of the behavior contract, which in this case is to help the student finish all of his assignments and hand them in when they are due. Emphasize that everyone must work together if the objectives of the contract are to be met.

- Discuss what the student must do that will help him improve his behavior. Asking the student for suggestions makes him a part of the process and demonstrates to him that he is responsible for his actions and has the ability to change them. For example, in the case of completing work, the student should be able to suggest things such as starting classwork immediately, not wasting time, and working the full amount of time provided. He should start homework when he comes home from school or right after dinner and work until he completes it. He should then place his homework in his binder.

- Set a time frame for evaluating progress. A weekly review, perhaps every Friday, provides enough time for progress to be made, yet is frequent enough that the student cannot become complacent and stop working toward improvement. Depending on the problem and the student, shorter or longer time frames may be set.

- Note any rewards and consequences. A reward might be a homework pass or the chance to participate in an after-school activity. Parents or guardians may also provide a reward, based upon a weekly progress report that you can e-mail them. Consequences might be after-school detention or a loss of privileges at home.

- Complete the contract. The student, his parents or guardians, and you should sign it.

There are many forms of behavior contracts. See the "Sample Behavior Contract" that follows, which you may modify to fit your specific needs. You may also search the Internet with the term "behavior contracts, students" for more examples. Before using any behavior contract, obtain the approval of your principal or supervisor.

Because they identify a specific problem and detail a plan the student can follow to overcome the problem, behavior contracts can be useful in managing inappropriate behaviors. However, contracts require patience and consistent effort from all involved. Behaviors do not change simply because they are explained on paper. Students need to work hard at changing their behavior, and you and their parents or guardians need to monitor, encourage, and support them through the process.

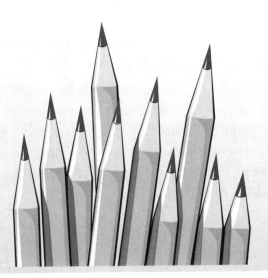

Sample Behavior Contract

Student's Name _____ Date _____ Period _____

Teacher's Name _____

Behavior to be changed: _____

_____

_____

To change this behavior, the student will: _____

_____

_____

To help the student change this behavior, the teacher will: _____

_____

_____

To help the student change this behavior, parents or guardians will: _____

_____

_____

Progress will be evaluated: _____

_____

Rewards for improved behavior: _____

_____

Consequences if behavior does not improve: _____

_____

The goals of this contract will be fulfilled when: _____

_____

Student's Signature: _____

Parent's or Guardian's Signature: _____

Teacher's Signature: _____

Involving Administrators in Addressing Inappropriate Behavior

Though minor behavior issues, such as chewing gum, occasional incomplete work, or daydreaming, do not require administrative involvement, serious problems, such as defiance, violent outbursts, or fighting, must be referred to an administrator.

In many schools, teachers are required to complete and submit a referral form to the office before an administrator will become involved with a behavior issue. The exception here is a sudden and serious situation, such as a fight between students, that must be handled immediately to ensure the safety of all students in the class. In such cases, you should contact the office for help. In other cases, a referral form provides an administrator with information that can assist him in addressing the problem effectively.

Following are steps to consider when referring behavior problems to an administrator:

- Make every effort to first solve the problem yourself. Address the behavior, speak with the student about appropriate behavior, take corrective action (perhaps moving the student's seat closer to your desk), speak with the student's parents or guardians about the behavior, and assign consequences—detention, for instance—in accordance with school policies.

- Speak informally with the administrator about the problem. She may be able to offer a suggestion that might help. This also makes the administrator aware of the problem. It is possible that the same student is misbehaving for other teachers, and a unified approach to solving the problem can be implemented.

- If the problem persists, complete the referral form. Be specific in describing the problem, list the steps you have taken to solve it, and include the results of your efforts. Remember that others, including the student's parents or guardians, may see the referral; be sure that it is professional in tone and language. (If your school does not require teachers to use referral forms, write down notes detailing the problem, your efforts to resolve it, and the results. When you meet with an administrator, refer to your notes to make sure that you do not overlook any important facts.)

- Keep a copy of the referral or your notes, as well as the date you submitted it to the office or spoke with an administrator.

- Avoid discussing the referral with the student during class or in front of others, which violates his privacy and may lead to a confrontation.

Once the administrator becomes involved, she will likely take the steps necessary to solve the problem. Be ready to provide any additional information or help she may request.

The Major Incident Report

Unfortunately, major incidents of inappropriate behavior, such as a fight, an uncontrollable verbal outburst, or apparent substance abuse, sometime occur in schools. Sudden, surprising, and serious, these incidents require that a student be removed from the class before he endangers or harms himself or others. You must act quickly. Contact the office for help in removing the student from the class. Never send students who are involved in a major incident to the office by themselves or simply order them to leave the room.

Because you must act so swiftly in these events, it may be difficult later to recall specific details of your actions, as well as the actions of others involved. However, an administrator will need clear details of the incident when he discusses the problem with the student and the student's parents or guardians. To ensure that you provide the administrator with accurate information, write down what happened immediately after the incident while facts are fresh in your mind. If your school has a form to record information about major behavior incidents, fill it out completely and submit it as soon as possible. If your school does not have a specific form for its teachers, use the "Major Incident Report Form" that follows. The form will serve as a record of the incident. Submit the completed form to your administrator and follow up the incident as necessary.

Major Incident Report Form

Teacher _____ Date of Incident _____

Place of Incident _____ Time _____

Student(s) _____

Description of Incident:

_____

_____

_____

_____

_____

Teacher's Action:

_____

_____

_____

_____

_____

Additional Comments:

_____

_____

_____

_____

Teacher Signature _____

Common Examples of Inappropriate Behavior and How to Handle Them

Throughout the course of any school year, you will contend with various behavior issues. Many will be easily resolved; a few will be ongoing. Though occasionally you may encounter an unusual or very serious problem, most of the problems you will need to manage will be relatively common to teachers across the country. Most can be managed by implementing specific strategies.

Following are several examples of inappropriate classroom behavior and suggestions for addressing them. If any inappropriate behavior continues after your initial efforts at management, you must contact parents or guardians and follow the guidelines in your faculty or school handbook. If necessary, refer problems to an administrator. (See "Involving Parents and Guardians in Addressing Inappropriate Behavior" and "Involving Administrators in Addressing Inappropriate Behavior" earlier in this section.)

HABITUAL LATENESS TO CLASS

Although most students who arrive late to class will have a valid excuse, you will no doubt encounter students who repeatedly arrive late for no apparent reason. These students may offer an explanation for being late, but repeated lateness becomes a serious disruption. Not only is the class activity interrupted, but by arriving late, students send a subtle message that your class is not important enough for them to be on time. Moreover, if their behavior is allowed to continue without consequence, other students may come to feel that they can be late, too. (See "Dealing with Students Who Arrive Late to Class" in Section Seven.) Address habitual lateness to class by doing the following:

- When late students enter the classroom, instruct them to sit down and start work. If necessary, briefly explain what they are to do. Avoid detailed explanation that will lengthen the interruption and risk losing the attention of other students.

- If you are already involved in the lesson, do not discuss with the student why he was late and why he should strive to be on time.

- After class, or after school, speak to the student about the importance of being on time to class. Find out if there is a legitimate reason for being late. For example, his previous class is at the other end of the school and that teacher does not let her students out at the bell. Take steps to resolve the problem. In this case, speak to the other teacher about dismissing students from class on time.

- Keep accurate records of tardiness. Do not give students a "break" when they are late as this will only encourage lateness. (Some schools have strict policies regarding lateness. Be sure to follow your school's guidelines.)

INATTENTIVENESS

Inattentiveness is a broad term with potentially significant implications for learning. Students who are inattentive may exhibit a variety of behaviors: boredom; disinterest in math; constantly finding reasons to get up out of their seats; daydreaming; staring out the window. Whatever the reason, their lack of focus on math and your class seriously undermines their progress. Address inattentiveness in these ways:

- Monitor the student closely. Whenever you sense she is wandering off topic, draw her back by asking a question or simply reminding her to refocus her efforts on math.

- As you circulate around the room, linger at this student's desk and encourage her in her work.

- Move the student's desk closer to yours or to the front of the room where you can monitor her more closely.

- Place this student in groups with motivated students who are likely to draw her into the activity.

- Differentiate your instruction to accommodate this student's particular learning style.

- If inattentiveness continues to be a problem, contact the student's parents or guardians. Be sure to rule out any medical conditions that might be causing the problem.

INCOMPLETE WORK

Students who do not complete work will not realize their full potential in your class. In addition, they are more likely to misbehave than other students, disrupting the class and disturbing others. Following are some ways to address the problem of incomplete work:

- Have clear rules regarding the completion of work, as well as consequences for incomplete work.

- Speak to the student about the importance of completing work on time. Explain that math skills are cumulative and build upon each other. Missing assignments makes learning new material more difficult.

- During class, remind the student to remain on task.

◎ Offer strategies for completing work at home. Students should start their homework right after school or after dinner, and continue working until they finish their assignments. They should begin long-term assignments well before the due date. You may also suggest that students work together in study groups, join homework clubs (if available), and participate in peer tutoring.

◎ Check the student's assignment pad or agenda to make certain that he writes his assignment down.

◎ Ask parents and guardians to check that the student completes his homework.

REPEATEDLY REQUESTING TO LEAVE CLASS

Some students regularly ask to leave class. They may ask to use the lavatory, see the school nurse, go to the guidance office, or go to their locker. In many cases, these students may be bored in class, not interested in math, or more interested in things outside of class. They may even have arranged to meet a friend outside of class. (See "Classroom Sign-Out Sheet" in Section Two and "Smoothly Handling Requests to Leave the Classroom" in Section Seven.) Address this problem in the following ways:

◎ Speak with the student about the importance of not missing class time. Do not excuse the student from incomplete work because she was out of class.

◎ Use a sign-out sheet to record when students are leaving your class.

◎ Review completed sign-out sheets to find out if a student is leaving class the same time each day. Although you may not refuse a student who claims she must go to the lavatory, you can ask her to remain until you are finished with the explanation of new material. However, do not refuse her request if she informs you that it is an emergency.

◎ If a student regularly asks to use the lavatory or see the nurse, check with the nurse about any health issues the student may have. If there are no physical disorders, speak with the student about her frequent requests to leave the classroom.

EXCESSIVE TALKING

Every class has a student or two (or more!) who are excessive talkers. Because these students will talk to anyone nearby, they can be a major source of classroom disruption, disturbing their peers and you. Address excessive talking by doing the following:

◎ Never ignore excessive talking in your class, and never try to teach when students are talking.

○ When students are talking, say, "Excuse me, but we need to focus on math." Such reminders are usually enough to bring students back on task.

○ Speak with the excessive talkers after class or after school. Explain that their talking unnecessarily is rude and disrespectful, disturbing others as well as undermining their own learning. Ask them what steps they can take to control their talking and help them formulate a plan. When students suggest ways they may control their talking, they assume at least part ownership of the problem, which may result in a prompt resolution.

○ Move the seats of excessive talkers close to your desk where you can monitor them easily. Be warned, however, as we have had the experience of moving excessive talkers next to our desks only to have them attempt to talk to us!

○ Place excessive talkers in groups where the other members are not likely to engage in off-task conversations.

○ Be consistent in enforcing classroom rules regarding unnecessary talking.

PASSING NOTES

When students are passing notes in class, their attention is not directed to math. Although note passing is not as disruptive as some behaviors, you should never ignore it. Address passing notes in these ways:

○ Confiscate the notes students are passing, but do not open or read them. Be aware that the contents of the note may contain personal or embarrassing information, which if you read may require additional action on your part.

○ If note passing becomes a chronic problem, inform the student to see you after school.

○ Discuss with the student that writing and passing notes during class prevents him from concentrating on learning math. Emphasize that you expect him not to pass any more notes in class.

○ Closely monitor note passers in your class. As you circulate around the room, be sure to stop by these students for a while, an action sure to deter note passing.

SLEEPING DURING CLASS

Various factors may cause a student to fall asleep in class. The student may be taking doctor-prescribed medicine, she may have a part-time job that prevents her from getting enough sleep, or she may be staying up too late at night. Maybe she has no interest in school and only reports to class because she must. When a student sleeps in your class, she certainly is not learning math and she is a distraction to others as they may become preoccupied in watching her nod off. Following are ideas to address this problem:

⊙ Do not permit a student to sleep in class. If the student is having trouble remaining awake, send her to the nurse.

⊙ Avoid creating a disturbance when waking the student. Walk to her desk and gently wake her.

⊙ Do not ask other students to wake a sleeping student. That is not their responsibility and usually elicits amusement, which compounds the disturbance.

⊙ Speak with the student after class or after school to find out why she is sleeping in class. Emphasize the importance of your math class.

  • If the cause is medication, perhaps the dosage needs to be adjusted. Or maybe the student needs to take the medication at a different time of day. Note: Medical issues and medications should be referred to the school nurse and the student's parents or guardians.

  • If she is sleeping because she works at a part-time job at night, suggest that she reduce her hours, if possible, or switch hours from evening to after school or to weekends.

  • If she is sleeping because she is up too late at night, encourage her to go to bed earlier.

  • If she is bored in your class, plan activities that appeal to her learning style and will interest her.

  • If you suspect that a student is depressed or suffering from a medical condition, refer her to the school nurse.

  • If you suspect substance abuse, refer her to the appropriate administrator.

  • If a student continues to sleep in class, contact her parents or guardians to discuss and resolve the problem.

ATTENTION SEEKING

Students may seek attention in numerous ways. They may comment on class happenings, joke, ask questions that are off topic, stroll to throw out trash in the midst of a lesson, make sounds, or simply act silly. These students will say and do outrageous things to be noticed. Their behavior is frustrating and disruptive to any learning environment. Address this behavior in the following ways:

⊙ Enforce appropriate class routines and procedures, and correct inappropriate behavior promptly.

- Avoid discussing the behavior during class, for this rewards the attention seeker. Require the student to see you after class or after school to discuss improving his behavior.

- Explain how his behavior is inappropriate for class and how it impairs his learning and that of other students. Emphasize that you value positive comments and positive participation in class activities, but cannot tolerate silliness or disruption.

- Assign tasks to the student where he can be noticed—for example, distributing materials or being a presenter for group work.

- Encourage the student to act in a positive manner and praise him when he does. Providing him with the attention he seeks for appropriate actions can reduce negative behavior.

INAPPROPRIATE USE OF TECHNOLOGY

The students in your class have a responsibility to use calculators, computers, and other technology in accordance with the guidelines established by you or your administration. Computers, for example, should be turned on and off correctly, used according to their purpose, and used only for class activities. Students should not use computers to access the Internet to visit unauthorized Web sites or check e-mail. Address inappropriate use of technology in these ways:

- Provide clear instructions for using technology in class.

- Give students reasons for specific procedures. For example, the reason for notifying you about any problems with equipment allows you to either fix the problem or promptly arrange for a technical person who can.

- Circulate around the room to provide assistance, as well as monitor equipment use.

- If you see students misusing technology, correct them and explain the potential consequences of their actions.

- In extreme cases, you may need to prohibit some students from using technology, or allow them to use it only under close supervision. Follow your school's policy for implementing consequences.

CELL PHONE USE

Most schools have clear policies regarding the use of cell phones by students. Despite these policies, many teachers still must deal with unauthorized cell phone use in their classes. The following examples occur in classrooms across the country: A student's phone rings during the explanation of a problem; a student on one side of the room sends a text message to her friend on the other side; a student at the back of class surfs the Internet with his phone nestled between the books piled on his desk. Cell phones can be a serious distraction in any class. Address the use of cell phones by doing the following:

- Enforce your school's policy regarding cell phone use by students. This may require students to turn off their phones in class and not take their phones out of their bags or binders. It may also require you to confiscate the phones of students who violate the policy.

- At the beginning of school, explain the school's policies regarding cell phones to your students. At back-to-school night explain the policies to parents and guardians. Emphasize that you must enforce the policies and that there can be no exceptions.

- If you must take a student's cell phone, keep it safe and hand it to an administrator as soon as possible. Be sure to fill out any required forms.

- Avoid using your cell phone in front of students, who will then question why you can use yours while they are not permitted to use theirs.

EATING OR DRINKING IN CLASS

There are many reasons why some students will bring food or drinks (water bottles, juice packets, or coffee) to class. They may be scheduled for the last lunch period and by your 11:00 algebra class may crave a snack; maybe they woke up late and did not eat breakfast; or maybe they want to test your reaction. Most schools have policies prohibiting, or severely restricting, eating and drinking in class, which can cause major distractions, be messy, and quickly evolve into a management headache as students munch on junk food during instruction. Spilled food or drink can also damage computers and calculators. Address this problem in these ways:

- Inform students of your school's policies regarding eating and drinking in the classroom, and enforce the policies.

- Be aware that students are likely to complain if not all teachers follow the policy. If they complain that Mr. Lee allows them to bring water bottles to class, calmly respond that you cannot speak for Mr. Lee, but that you are following school policy. Note, however, that if many teachers permit students to bring food and drinks to class, teachers who do not do so may be cast as unreasonable and unfair. If this situation occurs, you could politely ask the other teachers to follow the policy. If necessary, however, the school's policy should be reevaluated.

- Avoid eating or drinking in front of your students during class if they are not permitted the privilege.

INAPPROPRIATE WORDS AND COMMENTS

Every teacher knows students who say inappropriate words and make inappropriate comments. These students may use profanities, make sexual references, or utter racial slurs. Their words are offensive and hurtful and always cause a disturbance in class. Address this problem by doing the following:

- React to the word or comment calmly. Explain to the student that her choice of words is offensive and not appropriate for the classroom (or, depending on what she said, probably anywhere else). When she speaks in that manner, she is being rude and disrespectful to others.

- Do not confront the student by questioning her values, background, or upbringing. This will almost certainly put her on the defensive, and close off any willingness on her part to refrain from using such language again.

- Be aware of the feelings of the other students, especially those who might be the target of the remarks. You must not allow any student to verbally harass others. (See "Preventing and Responding to Bullying" in Section Eight.)

- Depending on the context and severity of the words, you may feel that it is necessary to refer the student to an administrator.

- Under no circumstances permit profanity or derogatory names and comments to be directed at you, students, or other adults.

DEFIANCE

Defiance can take many forms, from a student not following your request that he pick up the papers on the floor around his desk to talking back, arguing with you, and outright refusing to do what you ask. You must deal with any form of defiance promptly and firmly to avoid the impression that you are not in control of the class, which can result in other students believing that they, too, need not follow instructions or requests. Address the problem of defiance in these ways:

- No matter the degree or form of defiance, respond calmly. If you respond with anger, you risk an argument and make any resolution more difficult to achieve.

- Defuse the situation by saying to the student: "I must have heard you incorrectly." Comments like this give the student a chance to rethink his actions and avoid a more serious confrontation.

- Do not argue. If a student tries to argue with you, take him into the hall where you can speak to him calmly. (Remember that you are still responsible for the other students and you should position yourself so that you can continue monitoring the classroom.)

○ Speak with the student after class or after school. Explain that you cannot permit him to defy you or disrupt the class. Ask if something is wrong and offer your help. Mention that you are willing to listen to him, but that he needs to speak to you in an acceptable manner. Point out that everyone in class must work together to learn math.

STEALING

The best way to manage the problem of stealing is to be proactive in preventing it. Encourage your students to keep track of their possessions and not leave valuable items unattended. You, too, should follow this advice. Never leave personal belongings such as a purse, keys, flash drive, calculator, or grade book on your desk, a table, windowsill, or any other open area. Any money you collect from students should be stored in a locked drawer or cabinet and handed into the office as quickly as possible. Do not permit students to use items on or in your desk, such as pens or your calculator, and always lock your classroom door when you leave. When you are confronted with the problem of stealing, do the following:

○ If an item that belongs to you or a student is missing, first make sure that it has not simply been misplaced. A student may be convinced that her calculator has been stolen because it is not in her backpack when in fact she left the calculator in her locker earlier in the day. Suggest that the student search the places where the item might be.

○ If an item of yours or one of your students is stolen, explain to the class that the item is missing and you would like their help in finding it. Use a tone without accusation and do not say it was stolen. Ask if anyone may have picked it up by mistake or borrowed it without asking. If it is not returned at that point, explain that if anyone finds and returns it later, you will appreciate their efforts and will not ask any questions. Should the item be returned, keep the matter private.

○ If the item is valuable, and it is not returned, contact an administrator. If possible, do not allow students to leave the room until he arrives, at which point he will assume management of the problem.

○ In cases where you catch a student stealing, you should follow your school's policy regarding theft, which likely will require you to refer the incident directly to an administrator.

CHEATING

Students cheat for various reasons. They may want to receive good grades, but feel they are incapable of earning them; they may not have studied for a test; or they may not have completed work and do not want to face the consequences of not handing the assignment in. Cheating can distort your grading system; it is particularly distressing when a cheater receives grades he has not earned. When other students know that a classmate is cheating, they will often become resentful of behavior they recognize is dishonest but which seemingly benefits the cheater. Following are some ways to address cheating:

- Follow and enforce your school's policy on cheating. Make sure that students understand the policy and realize that cheating is dishonest.

- Explain to your students that cheating involves not only the student who is copying answers, but also any student who lets him copy. Both are being dishonest.

- Monitor students closely during tests and quizzes to discourage cheating. Circulate around the room and linger near any students you suspect might be cheating. Modify tests and quizzes slightly to prevent your morning students from passing answers to subsequent classes. You may even change a few problems or the order of problems slightly, then distribute the different versions to alternate rows or students. For homework and classwork, require students to show their work and not hand in papers with only answers.

- Never accuse a student of cheating without proof.

- Avoid accusing a student of cheating in front of the class, which usually results in a strong denial.

- Speak with the student after class or after school. Explain your suspicions and show the proof you have.

- Encourage students to work hard, be prepared for class each day, and take pride in their efforts for achievement.

VANDALISM

Vandalism is a recurrent problem in many schools. Typical acts of vandalism in the classroom include writing on desks, on walls, and in books; tearing pages from books; breaking off pencils in sharpeners; snapping off keys from computer keyboards; and stealing batteries from calculators. You no doubt can relate many more examples. Because it defaces or ruins school property, undermines pride in the school, and can disrupt classes, vandalism should never be ignored. Address vandalism by doing the following:

- Inform your students of your school's policy and explain that you must enforce the policy.

- If you find an example of major vandalism—for example, graffiti on the corridor wall—inform an administer as soon as possible. Note: Graffiti that refers to serious topics such as drug use, suicide, racial slurs, or violence should be reported immediately.

- Make a habit of visually scanning your classroom when you enter and just before you leave, paying particular attention to the desks, equipment, and walls. This is especially important if other classes use the room. Also, instruct your students to tell you at the start of class if the desk at which they sit has been vandalized. You can then check with the teacher of the previous class in an attempt to identify the student responsible for the vandalism. (If every teacher follows the same procedures, vandalism in classrooms can be substantially reduced.)

- When you find that a student has vandalized school property, speak to him about the importance of treating school property and materials properly. Follow your school's policies regarding discipline and restitution.

VERBAL ABUSE

When a student experiences severe frustration, she may lose control and become verbally abusive. She may lash out at other students, you, or other teachers. This can become a serious situation. Address verbal abuse in these ways:

- If possible, remove the student from the classroom. (Perhaps take her into the hall, but stand in a manner that you may still observe your classroom. Remember that you are still legally responsible for the students inside.)

- Allow the student to express her feelings and ask why she is upset.

- Explain that verbally abusive behavior is unacceptable.

- If you fear that she may become violent, do not permit her to stand too close to you.

- Allow her to reenter your classroom only after she has regained control of herself.

- If necessary, call the office or send another student for an administrator for assistance.

- If necessary, complete a "Major Incident Report From," presented earlier in this section.

VIOLENCE

Sometimes severe frustration can cause a student to experience a violent outburst. The violence may be directed at another student or students, property, or at you and other teachers. Any violent act requires that you act quickly. Address violence by doing the following:

- Immediately call the office or send a student for an administrator or another teacher for assistance.

- Try calming the student or students involved.

- Make sure that other students remain in their seats and safely out of the way.

- Follow your school policy. Many schools discourage teachers from attempting to physically stop or stand between students who are fighting. (Should a student claim he was hurt because of your actions, you can be liable.)

- Call or send a student for the school nurse if students are injured.

- Complete a "Major Incident Report Form," presented earlier in this section.

Your job would be much easier if every student behaved appropriately all the time. Unfortunately, that will never happen. Always use common sense and your school's policies as a guide when addressing inappropriate behavior. Although you will be able to solve many of the behavior problems you will encounter, some will require that you seek help from parents or guardians, administrators, or guidance counselors.

Quick Review for Managing Inappropriate Behavior

Inappropriate behavior of students has a negative impact on the educational setting of any classroom. To ensure that your students focus their energy and thoughts on learning math, you must establish practical classroom rules and promptly address instances of misbehavior.

The following suggestions can help you manage inappropriate behavior and create a classroom in which learning is the priority:

- Base the rules of your classroom upon your school's faculty handbook, discipline policies, and the expectations you have for your students.

- At the beginning of the year, inform your students of your rules regarding proper conduct and the consequences should they ignore the rules.

- Be consistent and fair in the enforcement of your rules and those of the school.

- Make every attempt to handle routine instances of inappropriate behavior yourself. Your students will view you as being in control of the class and are more likely to conduct themselves in an acceptable manner.

- Address instances of inappropriate behavior positively and professionally. Never lose your temper with students, never allow them to negotiate or argue with you, and never embarrass them by reprimanding them in front of others. Such actions will undermine your authority and diminish the respect students have for you.

- Do not hesitate to contact parents or guardians to enlist their help in resolving behavior problems with their children. When speaking with parents or guardians, explain the problem clearly, what you did to resolve it, the result of your efforts, and how they can help.

- Consider using behavior contracts with students to clarify problems and outline the steps students can take to improve their behavior.

- Follow your school's policies for involving administrators to help you manage serious or persistent student behavior issues.

- Remember that all teachers must contend with inappropriate behavior. Addressing misbehavior in a prompt, practical, and consistent manner is the most effective way to resolve behavior issues.

When you effectively manage inappropriate behavior, you are showing your students that you are in charge of the classroom. They will come to respect your authority and will be more likely to behave according to your expectations and the classroom's rules. They will view your class as a place to learn and enjoy math.

SECTION FOURTEEN

Working with Parents and Guardians

As teachers, we realize the importance of the involvement of parents and guardians in their children's education. Though many parents and guardians are in fact involved, others may want to help their children in school but not know how, and still others may be struggling with problems that prevent them from helping. Whatever the situation or background of parents or guardians, you need to encourage them to assume an active role in their children's schooling.

Parents and guardians can be your greatest resources in ensuring that your students reach their fullest potential. Their support of you and your math program, and of education in general, has a significant influence on their children's attitudes toward learning. Students who know that their teachers and parents or guardians are in agreement about learning usually work harder in school than students who feel that the adults in their lives have little interest in their success.

You must build positive relationships with your students' parents and guardians. By maintaining open lines of communication, keeping parents and guardians informed of the progress of their children, providing them with information how they can help their children do well in your class, and responding to their questions and concerns promptly and professionally, you will gain their support in the education of their children. Not only will this help your students enjoy greater success in your class, it will make your job easier and more satisfying.

The Expectations of Parents and Guardians for Their Child's Math Teacher

Although every school year you will have contact with a few parents or guardians who expect you to treat their child with more consideration than any other, most will have reasonable expectations for you and your math program.

Most parents and guardians expect you to:

- Act professionally with their child and them.
- Treat their child fairly.
- Provide a safe and enjoyable learning environment.
- Inform them about problems promptly.
- Inform them of your grading system and classroom policies.
- Teach their child the mathematical concepts and skills necessary for success in your class and as a foundation for success in the next grade and beyond.
- Utilize calculators and computers in class so that their child will have practical experience with technology.
- Provide experiences in critical thinking and problem solving.
- Prepare their child for standardized testing.
- Offer them suggestions for ways they can help their child succeed at math.
- Provide clear directions for assignments so that they may easily help their child with homework.
- Instill in their child an appreciation of the importance of mathematics to life.
- Encourage their child to do the best that he or she can in your class.

Most parents and guardians want to support their child's teachers. Being aware of what the parents and guardians of your students expect from you and your math program enables you to address their concerns and develop a positive relationship with them.

How to Make Parents and Guardians Partners in Math Education

The great majority of parents and guardians want their children to do well in school and are willing to support your efforts in the classroom. Though some will step forward and ask you how they can help their children achieve success in your class, others, because they may not have had good experiences during their own schooling or they are not sure of what they can do to help, will need encouragement. There is much you can do to welcome parents and guardians into a partnership for the educational benefit of their children.

The following suggestions can help you gain the support of parents and guardians:

- At the beginning of the year, inform them of your course description, classroom rules, and grading policies. (See "Rules and Requirements of Your Classes" in Section Two and "Devising a Fair System of Grading" in Section Twelve.)

- Invite and welcome parents and guardians to your classroom during back-to-school night where you can present your math program to them. (See "Preparing for and Conducting a Successful Back-to-School Night" later in this section.)

- Always treat them professionally and with respect.

- Contact them by phone or e-mail as soon as possible should you suspect a potential problem. (See "Involving Parents and Guardians in Addressing Inappropriate Behavior" in Section Thirteen.)

- Return phone calls and respond to e-mails from parents and guardians promptly. Try to answer any questions or concerns they may have.

- If your school has a homework hotline or maintains a Web site on which teachers can post assignments and classroom news, update your data regularly for parents and guardians so that they can work with their children at home.

- Consider writing a class math newsletter that highlights the work your students are doing and announces special class events. A newsletter need not be complex to be effective. You may post a newsletter on your school's Web site, send it as an attachment via e-mail to parents and guardians, or produce it on a single sheet of paper. (Placing information on both sides of one sheet can result in an informative newsletter.) Having students write the newsletter with you serving as editor makes this a truly worthwhile activity.

- Inform them of any math Web sites the class is viewing, particularly those that provide games and activities that reinforce skills.

- Encourage parents and guardians to attend conferences. (See "Conducting Successful Conferences with Parents and Guardians" later in this section.)

- Send home positive messages (e-mail is great for this!) when students do well in your class. Parents and guardians, and students, are always pleased to hear good news. Messages need not be long or overly detailed. Something like "Dear Mrs. Santos, I just wanted to let you know that Maria did a great job on her math project, earning an A+. I'm very pleased with her work."

- When you speak with parents and guardians, be patient and considerate, even if they are less cooperative than you feel they should be.

- Always keep any communication with them focused on their child. Never talk about other students, other parents or guardians, or administrators or teachers.

- Follow through with any actions you tell parents or guardians you will do. Never promise things you cannot accomplish.

- Always be friendly and thank parents and guardians for their support of their children, you, and your math program.

The time you spend helping parents and guardians become involved in their children's education will be well spent. Students will benefit from the support of their parents and guardians, and parents and guardians will realize the satisfaction of taking an active role in their children's schooling.

How Parents and Guardians Can Help Their Children with Math

Most parents and guardians understand the importance of mathematics in life, and they are willing to help their children acquire essential math skills. "Guidelines for Helping Your Child with Math," which follows, suggests actions they can take. You might reproduce this sheet and mail it to parents or guardians before school starts or hand it out at back-to-school night.

When parents and guardians become involved in the learning process, they can better understand your program and the way you present the course material. They will have fewer questions and concerns. Should they have to contact you, they will be able to discuss their child's success from a personal yet knowledgeable perspective.

Guidelines for Helping Your Child with Math

There are many ways you can help your child learn math. Following are some of the most important:

- Be willing to work with your child's teacher. Support the teacher's efforts by monitoring your child's homework and study habits. Make sure he or she completes assignments and studies for tests. Ask the teacher for supplemental materials if your child appears to be struggling.

- Speak positively about math and its importance. Never tell your child that you did not do well in math when you were young, or that boys are better at math than girls, or that math is difficult. Emphasize that people learn math by working hard.

- Familiarize yourself with your child's math program. Understanding the major topics, sequence, and format will enable you to provide help. Never express your dissatisfaction with the program or textbook to your child as this only provides your child with an excuse for not working hard.

- Provide the appropriate materials and supplies. If you are not sure what your child needs for math class, contact his or her teacher.

- Work with your child when he or she is having trouble with concepts or skills. Show patience and perseverance. When parents take time to help children with math, the children realize that math is important.

- Encourage your child to carefully read word problems. He or she should restate the problem, identify what the question is asking, and find the information needed to solve the problem.

- Encourage problem solving by asking questions such as:
  - What are you trying to find?
  - What information do you need to solve this problem?
  - How can you find the necessary information?
  - How will you organize your information?
  - Can you make a model or a sketch to help you find the solution?
  - Have you solved a similar problem? Can that experience help you solve this one?
  - Are there any patterns or relationships that can help you find the solution?
  - How do you know your solution is correct?
  - Are there other possible solutions?

- Ask your child to explain how he or she solves a problem.

- Encourage your child to double-check work, consider if an answer is reasonable, and correct any mistakes.

- Point out examples of how we use math every day—for example, reading recipes and measuring the amounts of ingredients when baking, interpreting statistics in sports, and using the scale on a map.

- Ask your child to help you solve math problems in real life, such as estimating the cost of items at a store, measuring the dimensions of a room for a new carpet, or calculating change.

- Obtain and play logic games and solve math puzzles.

Preparing for and Conducting a Successful Back-to-School Night

Back-to-school night (or open house in many schools) is a time you will meet many of the parents and guardians of your students for the first time. It is an opportunity for you to introduce yourself, present your course, and begin building a positive relationship with them. An informative and effective back-to-school night can help you enlist the support of parents and guardians for your math program.

A successful back-to-school night is a result of preparation and implementation. Begin preparing your presentation (or presentations if you teach more than one course) several days in advance by doing the following:

- Know how long your presentation should be. You do not want to finish too early, nor do you want to rush through the allotted time. In either case, parents and guardians may leave feeling that your presentation was poorly planned, which will reflect negatively on you as a professional.

- Plan what you will say. Your presentation should include a brief overview of your program, the math standards applicable to your course, important topics you will cover in class, how students will use technology, and major projects or events that will occur. The presentation should also include your policies, classroom rules, expectations, and grading system.

- Prepare handouts that detail pertinent information about your class. Your contact information—school phone number and school e-mail—should be included. (See the "Sample Back-to-School Night Information Sheet" that follows.)

- Consider using PowerPoint to create your presentation, which will enable you to easily highlight important information.

- Make sure that your classroom is clean and attractive. Display your students' work, hang posters that emphasize mathematical concepts, and exhibit textbooks and other materials the students will be using.

- Practice your presentation to make sure your delivery is smooth and fits your time frame.

To implement a successful back-to-school night, do the following:

- Dress professionally. Arrive at your classroom early and make certain that handouts are ready and that any equipment you will be using is working.

- Write your name and the name of your course on the board.

- Greet parents and guardians at the door. Introduce yourself and ask them to identify themselves. Distribute any handouts (doing so later wastes time during the presentation).

- Ask parents and guardians to sign in so that you will have a record of who attended. Have a sign-in sheet (with a pen) on a desk near the door. You may use the sign-in sheet that follows, or create one of your own. Asking parents and guardians to write their name as well as the name of their child eliminates confusion if last names are different.

- Begin your presentation promptly. Welcome any late arrivals warmly and continue the presentation. Do not restart the presentation for you will risk running out of time at the end.

- Speak with clarity and enthusiasm. Be personable and positive. Stand straight and make eye contact.

- Avoid speaking with any parent or guardian about his or her child. Back-to-school night is not a time for individual conferences. (Note: Some teachers purposely leave their grade books at home on back-to-school night so that they cannot speak about individual grades.) If necessary, offer to call the next day to discuss questions about individual students.

- Close your presentation by thanking those who came and telling them that you are looking forward to working with them through the coming year. Encourage them to contact you in the event a problem should arise or if they have any questions as the year progresses.

- If time is left, take questions; however, avoid answering any for which you are unprepared or any that are about individual students. Offer to call the next day when you will have time to discuss the question in detail.

Back-to-school night is your chance to formally present your math class and make a positive impression on parents and guardians. It enables you to welcome them into your classroom and encourage them to support their children in learning math.

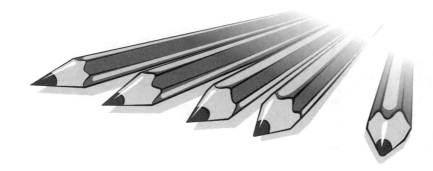

Sample Back-to-School Night Information Sheet

Welcome to Mr. Park's Algebra II Class

Algebra II provides the foundation for advanced study in mathematics. Major topics include systems of linear equations and inequalities, matrices, quadratic functions, polynomials, rational expressions, irrational and complex numbers, and probability. Students will learn to organize, analyze, and interpret data. They will also learn a variety of problem-solving strategies.

Our textbook, *Algebra II: Theory and Practice*, aligns with the current state math standards. Students are encouraged to visit the Web sites that are noted throughout the text and access the online guide for homework support, career links, application links, and data updates.

Each student has been assigned a TI-84 graphing calculator to use in the class and is responsible for returning the calculator at the end of class each day. Should the calculator be lost or damaged, a replacement fee will be charged.

Basic Rules for the Class

1. Students should be on time for class.

2. Students should be prepared. They should bring their textbook, assignment pad, two pencils, math notebook, and homework to class each day.

3. Students are responsible for making up the work they miss during absences. Makeup work is due on the day after students return to school. (For extended absences, students should see me about making up their work.)

4. Students should respect themselves, others in the class, and property.

Sample Back-to-School Night Information Sheet (continued)

Assessment

1. Tests count as 30% of the marking period grade. There are about three tests each marking period. They will be announced a week before they are given and usually coincide with the ending of a chapter in the text (although some long chapters will have two tests). Students should study for math tests by reviewing their class notes, homework, classwork, and the chapter review in their text.

2. Quizzes count as 20%. Quizzes are frequent and may be given at any time. The best way to be ready for quizzes is to review notes and homework each night.

3. Classwork counts as 15%. Answering (and asking) questions, completing problems on the board, making good use of class time, completing activities, and assuming a positive role in group work will enhance the classwork grade.

4. Homework counts 15%. Homework will be assigned each night except Fridays and before holidays or vacations.

5. Projects count 10%. One project will be assigned each marking period.

6. Math notebooks count 10%. Students should keep an assignment pad, class notes, classwork assignments, homework assignments, quizzes, and tests in their notebooks.

Contact Information

1. Contact me at school by phone at 123–456–7890, ext. 21.

2. Contact me via e-mail at jpark@highschool.org.

I try to respond to messages within twenty-four hours. Please be sure to leave a phone number and a time I may reach you.

Best wishes for a great year!
Mr. Park

Parent-Guardian Sign-In Sheet

Teacher's Name _____ Date _____

Course _____ Period _____

Parent's or Guardian's Name	Student's Name

Conducting Successful Conferences with Parents and Guardians

Parent-teacher conferences provide an opportunity for you to meet with parents and guardians to discuss the progress of their children. During conferences you may offer suggestions for how they can help their children succeed in your class, present strategies for resolving problems their children may be experiencing, and discuss any concerns you or they may have. Whether conducted during the time your school reserves for conferences, or at some other time during the year, an effective conference is beneficial to students, their parents or guardians, and you.

Following are several steps you can take to ensure that the conferences with your students' parents and guardians are successful:

- Before the conference, review the student's work and grades. Identify a goal for the conference. Maybe the student needs to be more attentive in class, complete homework consistently, or improve his test scores through studying. Perhaps he needs to be more respectful of other students, or to come to class on time each day. Maybe a good student simply needs to keep doing what he is doing.

- Make notes of the major points you would like to address during the conference. Include the student's strengths and weaknesses; however, keep in mind that if your conferences are limited by time you must prioritize the points you would like to make. If you need more time to discuss serious issues, schedule another conference at a later date.

- Gather samples of the student's work to show his parents or guardians. Consider placing notes and the work of each student in a separate folder.

- Think about the questions parents or guardians might ask you about their children. For example, if you are using a new math program, it is likely they will ask about the text. Perhaps they will have questions about the use of technology in your class, or they may wonder why another math teacher gives less homework than you do. Anticipating questions reduces the chances that you will be surprised and provide an immediate response that you later wish you could retrieve. (Note: Politely refrain from discussing another teacher's program or class.)

- Invite any in-class support teachers to conferences with the parents or guardians of students for whom they are responsible.

- Arrange furniture to make a comfortable setting, such as by placing chairs around a table. Do not sit behind your desk with the parent or guardian in front of you as this can give the impression that you are aloof or distant.

- Have paper and pens available for you and for parents and guardians to write down notes.

○ Begin the conference promptly. Express pleasure for meeting the parent or guardian. Be friendly and welcoming, but professional. Be sure to offer some positive remarks about the student.

○ If a student is having problems, focus the conference by clearly stating your concerns. Be specific and offer facts to support your statements. Offer examples of work that illustrate academic problems, and provide instances and details of inappropriate behavior. Explain the steps you have taken to solve the problems and the results of your actions.

○ Develop strategies to resolve the problems the student is experiencing. Clearly establish what the parents or guardians should do at home and what you will do in class.

○ Be a good listener. Sometimes parents or guardians are desperate to help their children but feel powerless.

○ Always remain calm and professional during discussions. Never put a parent or guardian on the defensive by implying that she is to blame for her child's problems. If a parent or guardian becomes upset, remain composed and remind her that both of you must work together to help the child.

○ Be ready for the parent or guardian who tries to shift the fault for his child's lack of progress on to you or the school. Such parents or guardians might say that their child is having trouble concentrating in class because the classroom is noisy, the textbook is too hard, or the curriculum is not meeting her needs. Respond by calmly refocusing the discussion on the student's academic strengths and weaknesses, behavior, and attitude.

○ Maintain detailed notes of "difficult" conferences. Although most conferences will be routine and result in your providing parents and guardians with simple suggestions for helping their children, some conferences will address serious problems. For these conferences, you should take notes during the conference, recording the problem, strategies that were discussed, and the actions that the parents or guardians and you will take to resolve the problem. Completing the following "Parent-Teacher Conference Log Sheet" results in a detailed record, which can be extremely helpful should additional action need to be taken.

○ Keep conferences focused on the child of the parent or guardian. Never discuss other students or adults.

○ At the conclusion of the conference, review the major points that were discussed and actions that are to be taken. If a follow-up is necessary, explain when you will contact the parent or guardian via e-mail or phone or when they should contact you. Be sure to follow through.

○ Thank the parent or guardian for coming to the conference.

At your parent-teacher conferences, parents and guardians will look to you to guide them toward ways they can support their children in school and particularly in your math class. When you approach conferences with a clear purpose and practical suggestions for the improvement of your students, you give parents and guardians the opportunity to become active participants in their children's math education.

Parent-Teacher Conference Log Sheet

Student _____ Period _____ Date of Conference _____

Parent or Guardian _____

E-Mail _____

Parent or Guardian Home Phone _____ Cell _____

Topic(s) of Concern: _____

_____

_____

_____

_____

Strategies Suggested: _____

_____

_____

_____

Strategies Parent or Guardian Will Implement: _____

_____

_____

_____

Strategies Teacher Will Implement: _____

_____

_____

_____

Follow-Up: _____

_____

_____

_____

Dealing with Difficult Parents and Guardians

Although the majority of your students' parents and guardians will be wonderful to work with, there will be some who will never be satisfied with your efforts on behalf of their children, or your math program, or sometimes the entire school. There are many kinds of difficult parents or guardians, from those who are uncooperative to those who are intensely critical of the education their child is receiving. It is not easy dealing with such parents and guardians.

The following strategies can help you work with difficult parents and guardians for the benefit of their children:

- Be professional at all times. Always remain calm; never lose your temper or raise your voice.

- Make it clear that you and they need to cooperate in the best interests of their child.

- Follow school rules and procedures.

- Keep accurate records of meetings, conferences, e-mails, and phone calls. (Use the "Parent-Teacher Conference Log Sheet," presented earlier in this section, or "Record of Parent-Guardian Contact" in Section Thirteen.)

- Be willing to listen to difficult parents and guardians. Refrain from interrupting them and allow them to fully explain their concerns. Interrupting them, even to correct a misinterpretation of a fact, may cause or worsen a confrontation. After they are done, you can present your side.

- Ask questions when you need clarification. Not only does this foster good communication, it can uncover and help resolve misunderstandings.

- Restate the problem to ensure that you understand the concerns of the parent or guardian.

- Explain your position clearly. Offer specific details and reasons for actions you have taken.

- Offer suggestions or strategies for resolving problems.

- If you are unable to resolve the problem, request the help of a guidance counselor or an administrator.

Difficult parents and guardians, like all others, want their children to do well in school. Convincing them that you share the same goal is an important step to establishing a working relationship with them. Once you gain their support, work together for the benefit of their children.

Working with Parents and Guardians Who Speak Limited English

Communicating with parents and guardians who speak limited English presents a special challenge. Although they want their children to be successful in school, their limited ability to converse in English impedes their involvement with their children's education. They are unable to offer much help with assignments, may not come to back-to-school night or conferences, and are sometimes difficult to reach via e-mail or telephone.

Most school districts have plans to assist teachers when they need to contact parents and guardians who possess limited English skills; however, managing effective communication remains the responsibility of the teacher. The following strategies can alleviate what otherwise can be a difficult problem:

- Follow your school's plan for working with parents and guardians who speak little or no English. This may involve checking with the student's guidance counselor, who may have helpful information. The guidance counselor may know of a staff member, trustworthy older sibling, family member, or community volunteer who can serve as a translator.

- Depending on the issue, if the student is reliable, mature enough, and speaks both languages capably, consider having her serve as a translator. A student should not serve as a translator for sensitive issues.

- If the student is in your school's ESL program, consult with her ESL teacher, who may serve as interpreter or may know of someone who can.

- Be aware of cultural differences. Parents or guardians will appreciate your efforts at understanding their customs.

- When communicating with the parent or guardian, speak slowly and clearly, and give the parent or guardian time to process your words. Try to avoid words and terms that are likely to be confusing, for instance, "75th percentile," "ESL," or "heterogeneous group." Also avoid idioms such as "it's raining cats and dogs," "let the cat out of the bag," or "once in a blue moon," which are taken for granted by native English speakers but can be confounding to people learning English. However, be careful that your attempts to use understandable language do not appear to be condescending.

- Send letters home in the native language. Perhaps your school's ESL teacher can help you.

- Use the Internet as a resource. Two user-friendly Web sites through which you can access electronic translators are http://babelfish.yahoo.com and www.ilovelanguages.com.

Working with parents and guardians who have limited ability speaking English requires patience and consideration. Imagine what it would be like if you moved to another country and could not speak the language yet had children enrolled in the local school system. The effort you make to communicate with these parents and guardians for the benefit of their children will be greatly appreciated. (See "Students Who Speak Little or No English" in Section Eight.)

Expanding Your Role as a Math Teacher

Today's truly effective teachers know that their role extends beyond the classroom. They realize that they are a vital part of an educational community that includes students, colleagues, administrators, support personnel, and parents and guardians. Their responsibilities are not limited to teaching.

Many teachers expand their roles by becoming involved in school activities outside their classrooms—coaching, advising for clubs, or chaperoning school events—but most of these activities focus on students. (See "After School and Beyond" in Section Four.) An excellent, though often overlooked, way for math teachers to expand their role is to offer math workshops for parents and guardians. Workshops that show them ways they can help their children succeed in your class and how they can foster in their children an appreciation of mathematics can build support for your program and enhance student learning. A productive math workshop for parents and guardians requires careful preparation, but its benefits can be significant.

Following are some important guidelines for planning:

- Formulate your basic idea for a workshop and discuss it with your principal or supervisor. Gaining his or her approval assures you of the necessary administrative support.

- Select a date and time for your workshop. Avoid scheduling it at the same time as another major school activity such as a sports event, concert, or drama production. Also avoid days before or after holidays or vacations. Scheduling your workshop on the night of a general PTA meeting can improve attendance.

- Identify your audience. Will you limit your workshop to the parents and guardians of your students? To all parents and guardians of the students of a specific grade level or course? Or to all parents and guardians of the students of the entire school? Remember that the bigger and more diverse your audience, the more difficult it will be to provide meaningful activities and information. The needs of a parent or guardian of a fifth-grade math student are different from those of an eighth-grade student.

- Determine the number of people you can accommodate at your workshop. If you are working alone, more than twenty-five will be hard to manage.

Working with a colleague will allow you to present your workshop to a larger group.

○ Decide on a topic, purpose, content, and activities for your workshop. There are many topics you may present, such as:

- Introduce parents and guardians to a new textbook, explaining how they can help their children use the text's components.

- Review and explain your teaching techniques and how parents and guardians can support your instructional program.

- Demonstrate the use of a calculator, providing a hands-on presentation of its menus and capabilities.

- Discuss the math standards that drive your curriculum and how they affect instruction and student learning.

- Discuss open-ended problems and provide examples of scoring rubrics.

- Explain how writing is essential to mathematics, offer examples of ways students write about math, and discuss how parents and guardians can encourage children to express mathematical ideas in writing.

- Provide a sample lesson that gives parents and guardians insight into your curriculum.

- Provide information about math assessments, including standardized tests and how students may prepare for tests. (A guidance counselor will be a great asset to this type of workshop.)

- Introduce and explore math Web sites for students and parents and guardians. (See "Resources on the Internet" in Section Three.)

○ Plan your workshop. Create materials, handouts, and displays. Obtain or reserve any special materials or equipment you will need.

○ Announce your workshop in district, school, and PTA publications, post information about it on your school's Web site, and send flyers home with students. If space is limited, include a sign-up slip on the bottom of flyers with a deadline.

○ Consider serving refreshments, such as cookies or small pastries, coffee, and tea. Be sure to keep receipts for reimbursement.

Thorough planning lays the foundation for an effective workshop. You can ensure that your workshop is successful by doing the following:

○ Arrive early and complete any final tasks. Make certain that you have all the necessary materials and equipment, and that enough desks (or tables) and chairs are set up. Place the refreshments at the back of the room. If a colleague is assisting you, review what each of you is to do.

◎ Make certain that all equipment, such as a projector, computer, and calculators, are working properly. If parents and guardians are to use calculators, have a few extra ones ready in case of malfunctions. Also have new batteries available.

◎ Greet parents and guardians at the door. Distributing any handouts as they enter the room eliminates the need for taking time to pass materials out during your presentation.

◎ Begin the workshop promptly at the designated time. Introduce yourself and explain the objectives of the workshop. If the group is small enough, ask parents and guardians to introduce themselves.

◎ Briefly explain how this workshop will help parents and guardians support their children's learning. From here, move directly into the activities of the workshop.

◎ Depending on the length of your workshop, you may want to provide a break of about ten minutes at the halfway point. As people help themselves to refreshments, circulate around the room and answer any questions they may have.

◎ At the conclusion of your workshop, summarize your main points and allow time for questions. You could hand out an evaluation form, asking parents to rate the different parts of your presentation and offer suggestions for how you might improve it.

You can extend a workshop for parents and guardians by including their children. In a family math workshop, you can provide hands-on activities that students and their parents or guardians work on together in a unique setting that fosters cooperation and support for learning math.

Ideally, a family math night should focus on the math that students are learning in your class. Include activities and problems, hands-on experiences with manipulatives that illustrate mathematical concepts, and ways that parents and guardians can support their children's efforts in your class. You could include technology, computers and calculators, for example, that students and their parents or guardians can use to solve specific problems.

A family math night is a time that parents and guardians can share some of the experiences of their children in your class. It is an excellent way for them to see evidence of their children's learning and for you to gain their support for your program. An Internet search with the term "family math night" will provide you with many helpful Web sites.

Quick Review for Working with Parents and Guardians

Parents and guardians have enormous influence over their children's attitudes toward school. Those who support education play a significant role in their children's progress. Working together, you and parents and guardians can provide students with the necessary environments—in class and at home—for learning mathematics.

The following tips can help you develop strong working relationships with the parents and guardians of your students:

- Be aware of what parents and guardians expect from you. The expectations of most are simple—that you act professionally at all times, treat their children fairly, and provide a safe and enjoyable environment for learning.

- Encourage parents and guardians to attend back-to-school night, school functions, and conferences.

- Keep parents and guardians informed of your class's assignments, activities, and special events. Make sure that all information is current on homework hotlines and in your postings on your school's Web site.

- Contact parents and guardians as soon as possible should you suspect a potential problem.

- Offer specific suggestions, strategies, and methods for helping parents and guardians help their children learn math.

- Be considerate and patient with parents and guardians, especially those who prove to be difficult to work with. Explain that you and they need to cooperate in the best interests of their child.

- Take the extra steps necessary to bring parents and guardians who speak limited or no English into the educational process so that they can assume active roles in their children's learning. Speak slowly and clearly, allow time for them to process your words and, if necessary, enlist the help of translators. Use the Internet as a resource for communication.

- Expand your role as a teacher by offering math workshops to parents and guardians.

When teachers, parents, and guardians are united in purpose and expectations, students receive a strong message of the importance of school. Students who see their parents or guardians make the effort to learn about their classes, attend school events, and keep in contact with their teachers tend to have positive attitudes about school and are diligent in their work. These students are more likely to succeed in math than students whose parents remain uninvolved and demonstrate little interest in their children's education.

SECTION FIFTEEN

Keeping the Flame Burning

Teaching is a demanding profession with daily challenges. Yet, for those who are able to handle the stress that comes with managing a classroom of students, it is among the most enjoyable and rewarding of careers.

As a group, teachers share many wonderful qualities. Most tend to be caring, conscientious, and considerate. They are committed to their students and dedicated to education. Because they give so much of their time for the benefit of others, however, they can easily overlook taking care of themselves. They can wear themselves out emotionally and physically, tarnishing the quality of their lives inside and outside of school.

All teachers must cope with stress. Although some stress is good for us, keeping us at our best, persistent stress that results in frustration, worry, and exhaustion steals our enthusiasm and energy. Most teachers who suffer extreme, long-term stress eventually leave the classroom; those who remain develop negative attitudes that deny them the delights of teaching and impede the positive learning experiences of their students. To be an effective teacher who finds satisfaction and enjoyment in your work, you must avoid burning out and instead keep the flame for teaching burning.

Causes and Symptoms of Teacher Burnout

Burnout is a state of emotional and physical fatigue brought on by continual and excessive stress. For math teachers, stress can arise from various factors and situations. Some of the most common include:

- The accountability for teaching a subject where student knowledge is assessed by various standardized tests
- The constant planning

- The daily demands to create interesting math lessons and provide effective instruction to address diverse learning styles
- The frustration of managing difficult students
- The frustration of working with uncooperative parents and guardians
- The need to meet numerous requests of administrators and supervisors
- The seemingly unrelenting paperwork, record keeping, and deadlines
- The feeling of being unappreciated
- The lack of adequate materials and equipment for teaching math, such as manipulatives, calculators, or computers
- The added pressure of supervising extracurricular activities

Burnout creeps up slowly. Minor frustrations with the daily routines of teaching build gradually, leading to exhaustion and a vague sense of dissatisfaction. These conditions continue to worsen until teaching is no longer enjoyable. Positive attitudes give way to negative ones that can spill over into other aspects of life.

Although every teacher reacts to burnout in his or her own way, common symptoms include the following:

- Lack of enthusiasm and energy; fatigue; a chronic feeling of being run-down
- Lack of interest in teaching; decreased motivation to do a good job
- Difficulty concentrating
- Little interest in creating innovative lessons
- A deep feeling of isolation
- Negative feelings or opinions about teaching or education
- Lack of confidence in the classroom
- A feeling of dread, anxiety, or apprehension when coming to school
- Cynicism, irritability, or impatience with students, colleagues, parents and guardians
- Apathy, arising from a sense that nothing you do matters
- A persistent feeling of being overwhelmed
- Feelings of helplessness, of being powerless, ineffective, or trapped
- A change in sleep patterns, especially increasing sleeplessness
- Exploding easily at minor problems
- Stomach or digestive ailments
- An inability to find satisfaction in what you do

Unlike the occasional and normal "down-in-the-dumps" feeling that everyone experiences from time to time and which disappears in a few days, the symptoms of burnout persist, often worsening. Fortunately, there are steps you can take to overcome burnout and, even better, ways to avoid it.

Avoiding and Overcoming Teacher Burnout

The key to avoiding and overcoming burnout is stress management. You must assume a proactive stance in dealing with the stress you face each day. By coping with stress effectively, you reduce its negative impact on you and prevent it from leading to burnout.

The following strategies can help you manage stress:

- Accept the challenge of teaching. Realize that stress is a daily part of the job and that it must be managed. Every teacher experiences stress; the best ones handle it effectively.

- Handle change gracefully. New students and new schedules each year, new courses to teach, new administrators with new ideas, new colleagues, new school policies, a transfer to a new grade or new school—teaching is full of changes. Reacting to change positively will help to make these new situations successful.

- Establish realistic goals for yourself and your students. Setting impossible goals will only result in frustration and stress.

- Establish practical rules for your classroom and enforce the rules consistently. Maintaining an orderly classroom minimizes stress.

- Set up a realistic schedule for yourself. Use your time in school efficiently—for example, by spending your free period planning and grading papers. Try to grade as many papers in school as you can and take home as little schoolwork as possible. Do not procrastinate.

- Plan and deliver a variety of interesting lessons. When students are engaged in a class's activities, they usually learn more and behave better. You will also experience satisfaction when you present innovative activities that excite your students about math.

- Arrive at school and at each class on time. Assigning a do-now to start class gets students working on math immediately and allows you a few moments to take attendance and organize your thoughts and materials for the coming lesson.

○ Be flexible. Understand that no matter how well planned or prepared you are, some days simply will not turn out as you expected.

○ View problems from a positive perspective and their solutions as a chance for everyone involved to gain new insight and understanding.

○ Update and revise your lesson plans from year to year. This will help keep your classes fresh and innovative.

○ Remain current with technology. As the capabilities of calculators and computers advance, incorporate the power of the advancements into your program.

○ Keep up with paperwork and record keeping. Attempting to grade dozens of papers and enter grades the night before they are due is a sure stressor. Use technology such as an electronic grade book to improve your efficiency.

○ Avoid assuming too many responsibilities. If you are feeling somewhat overwhelmed with your teaching schedule, do not volunteer to chair your school's social committee.

○ Seek out and interact with coworkers who are positive about teaching, their students, and your school. Limit the time you spend with the complainers on your staff. Overly negative people can influence you to view teaching with pessimism.

○ Work collaboratively. Working with a colleague can bring out the best in both of you. Each of you will gain fresh perspectives, approaches, and ideas.

○ Teach math at a different grade level or teach a new course to prevent yourself from becoming complacent. Teaching a new course will help keep your mind sharp by challenging you with new content.

○ Grow professionally. Join professional organizations; take some graduate classes; attend seminars, workshops, and conferences; visit math and educational Web sites; and read educational periodicals. (See "Maintaining Your Professional Expertise" in Section One, "Resources on the Internet" in Section Three, and "References and Suggested Reading" at the end of this book.)

○ Do not neglect yourself. Eat a healthy diet, get a sufficient amount of sleep, and follow an exercise program. (Note: Check with your doctor before beginning any exercise program.) To avoid every cold your students have, keep tissues and hand sanitizers available and wash your hands often. Teachers who maintain a healthy lifestyle have more energy and brighter outlooks than those who have allowed themselves to slip into poor health habits.

○ Take time for yourself. Remember that teaching is your job, and that you have a life outside the classroom. Stimulate your mind and relax your spirit by pursuing a new hobby or trying a new activity. Make sure you set aside time

for personal obligations and for relaxing with family and friends. When you are home, do not allow yourself to worry about your responsibilities in school.
- Be willing to laugh. Humor is a great tension reliever.

You cannot be an effective teacher if you are under excessive amounts of stress. Incorporating these strategies into your life will help you manage stress, avoid burnout, and allow you to enjoy the many satisfactions of teaching.

Becoming the Most Effective Math Teacher You Can Be

To be the best math teacher you can be, you must be willing to critically evaluate not only your performance in the classroom, but also your work as a member of your staff. You must continually grow as a professional in all aspects of your career.

Though the observations of administrators and supervisors can be sources of information regarding your skills in class (see "Evaluations for Math Teachers" in Section Four), so can the observations of your students, though in an informal manner. When asked their opinions about a class, most students can be quite honest and offer useful insights and suggestions. You can solicit student evaluations of your class by asking them to complete the "Student Course Evaluation Form" that follows.

Student Course Evaluation Form

Name (Optional) _____ Date _____

Course _____ Period _____

Please complete this evaluation. For numbers 1–12, circle the number that best describes your opinion of each statement about this class. Provide short answers for numbers 13 and 14.

	Agree				Disagree
1. Lessons and activities were interesting.	5	4	3	2	1
2. Directions for assignments were clear.	5	4	3	2	1
3. The teacher understood the subject.	5	4	3	2	1
4. The teacher made math easy for me to learn.	5	4	3	2	1
5. Students were encouraged to share ideas.	5	4	3	2	1
6. I was not afraid to ask questions.	5	4	3	2	1
7. The class was orderly.	5	4	3	2	1
8. The teacher gave extra help when students needed it.	5	4	3	2	1
9. The teacher treated everyone fairly.	5	4	3	2	1
10. Grading was fair.	5	4	3	2	1
11. I enjoyed this class.	5	4	3	2	1
12. I learned a lot of math this year.	5	4	3	2	1

13. The topic(s) or activity I enjoyed most this year was

_____

_____

14. This class can be improved by

_____

_____

_____

_____

The end of the school year is also the time to reflect upon the year and self-evaluate your performance. Answering the following questions is a good start for sharpening your professional skills:

- Am I completely familiar with the content and subject matter of my curriculum?
- Does my mathematical knowledge extend beyond my curriculum?
- Am I constantly striving to improve my teaching?
- Am I a capable facilitator?
- Do I use the Standards of the NCTM, state standards, and district goals in planning my lessons?
- Is my instruction clear?
- Do I integrate technology in my instruction and student activities?
- Do I provide a variety of activities that address the needs of students of various learning styles?
- Do I meet the needs of all my students?
- Do I plan activities that encourage my students to work cooperatively?
- Do I pose problems that relate to my students' lives?
- Do I foster critical thinking?
- Do I teach and encourage the use of various strategies for problem solving?
- Do I differentiate my instruction so that all students can be successful?
- Do I encourage the use of technology in problem solving?
- Do I maintain an orderly classroom with practical procedures and routines?
- Am I consistent in enforcing classroom rules and policies?
- Do I use a variety of ways to assess student learning?
- Am I fair and consistent in grading my students?
- Do I return student papers promptly?
- Do I communicate with parents and guardians as often as needed?
- Do I encourage parents and guardians to become partners with me in the math education of their children?
- Do I work well with colleagues, administrators, and support staff?
- Do I meet deadlines for completing paperwork?
- Do I attend in-services, conferences, and workshops that foster my professional growth?
- Do I read articles and periodicals about the best practices in mathematics education?

- Am I open-minded and willing to see various sides of issues?
- Am I a good role model for my students?
- Do I bring out the best in all my students?

Your teaching career is constantly evolving. Each year you will learn something new, try a new technique or method, attempt to solve a problem in a new way, or utilize new resources. Use the answers to the previous questions to identify your strengths as a teacher and the areas in which you need improvement. By building on your strengths and working to improve any weaknesses, you will grow in your career as a teacher of mathematics.

Quick Review for Keeping the Flame Burning

When you decided to become a math teacher, you chose a profession that comes with great responsibilities. Along with the typical rigors of teaching—classroom management, working with students who possess various abilities and personalities, paperwork and record keeping—you must plan lessons that address diverse learning styles, deliver effective instruction, and ensure that your students meet state standards and district goals. Another significant, though often overlooked, responsibility is the need to keep your enthusiasm for teaching high. You must keep the flame for teaching burning.

Consider the following tips:

- Understand that stress affects all teachers and can erode enthusiasm.
- Know that to be an effective teacher you must be able to manage stress.
- Realize that long-term, excessive stress can lead to burnout.
- Learn to recognize the causes of stress and symptoms of burnout.
- Take positive steps to avoid and overcome burnout.
- Use the information you gain from evaluations by your supervisors, course evaluations by your students, and self-evaluations to critically assess your effectiveness in the classroom and your position as a member of your school's faculty.

When you keep the flame for teaching burning, you maintain the mind-set and motivation for improving your professional skills and becoming the best math teacher you can be. You will treat each day as a challenge, not an obstacle. You will provide your students with quality instruction and be a positive factor in your school community. Every day you will renew your commitment for sharing a productive and rewarding year with your students.

Best wishes for your continued success.

References and Suggested Reading

Allen, Barbara, and Sue Johnston-Wilder (eds.). *Mathematics Education: Exploring the Culture of Learning*. New York: Routledge Falmer, 2004.

Allsopp, David H., Maggie M. Kyger, and Lou Ann H. Lovin. *Teaching Mathematics Meaningfully: Solutions for Reaching Struggling Learners*. Baltimore: Paul H. Brookes, 2007.

Brumbaugh, Douglas K., Enrique Ortiz, and Regina Harwood Gresham. *Teaching Middle School Mathematics*. Mahwah, NJ: Erlbaum, 2006.

Burns, Marilyn. *About Teaching Mathematics: A K–8 Resource* (3rd ed.). Sausalito, CA: Math Solutions, 2007.

Clapham, Christopher, and James Nicholson. *The Concise Oxford Dictionary of Mathematics* (3rd ed.). New York: Oxford University Press, 2005.

Courant, Richard, Herbert Robins, and Ian Stewart. *What Is Mathematics? An Elementary Approach to Ideas and Methods*. New York: Oxford, 1996.

Curriculum Focal Points for Prekindergarten Through Grade 8 Mathematics: A Quest for Coherence. Reston, VA: National Council of Teachers of Mathematics, 2006.

English, Lyn D., and Graeme S. Halford. *Mathematics Education: Models and Processes*. Mahwah, NJ: Erlbaum, 1995.

Fennema, Elizabeth, and Thomas A. Romberg (eds.). *Mathematics Classrooms That Promote Understanding*. Mahwah, NJ: Erlbaum, 1999.

Focus in High School Mathematics: Reasoning and Sense Making. Reston, VA: National Council of Teachers of Mathematics, 2009.

Forsten, Char, Jim Grant, and Betty Hollas. *Differentiated Instruction: Different Strategies for Different Learners*. Peterborough, NJ: Crystal Springs Books, 2002.

Franceschetti, Donald R. (ed.). *Biographical Encyclopedia of Mathematicians*. New York: Cavendish, 1999.

Gullberg, Jan. *Mathematics: From the Birth of Numbers*. New York: Norton, 1997.

Gurganus, Susan Perry. *Math Instruction for Students with Learning Problems*. Boston: Allyn & Bacon, 2006.

Huetinick, Linda, and Sara N. Munshin. *Teaching Mathematics in the 21st Century: Methods and Activities for Grades 6–12* (3rd ed.). Upper Saddle River, NJ: Prentice Hall, 2007.

Johnson, Art V. *Teaching Mathematics to Culturally and Linguistically Diverse Learners*. Boston: Allyn & Bacon, 2009.

Mathematics Assessment: A Practical Handbook for Grades 9–12. Reston, VA: National Council of Teachers of Mathematics, 1999.

Mathematics Assessment: A Practical Handbook for Grades 6–8. Reston, VA: National Council of Teachers of Mathematics, 2000.

Midkiff, Ruby Bostick, and Rebecca Davis Thomasson. *A Practical Approach to Using Learning Styles in Math Instruction*. Springfield, IL: Thomas, 1996.

Montague, Marjorie, and Asha K. Jitendra (eds.). *Teaching Mathematics to Middle School Students with Learning Difficulties (What Works for Special-Needs Learners)*. New York: Guilford Press, 2006.

Muschla, Judith A., and Gary Robert Muschla. *Math Starters! 5- to 10-Minute Activities That Make Kids Think, Grades 6–12*. San Francisco: Jossey-Bass, 1999.

Muschla, Judith A., and Gary Robert Muschla. *The Math Teacher's Book of Lists, Grades 5–12* (2nd ed.). San Francisco: Jossey-Bass, 2005.

Muschla, Judith A., and Gary Robert Muschla. *Hands-on Math Projects with Real-Life Applications, Grades 6–12* (2nd ed.). San Francisco: Jossey-Bass, 2006.

Muschla, Judith A., and Gary Robert Muschla. *The Math Teacher's Problem-a-Day, Grades 4–8*. San Francisco: Jossey-Bass, 2008.

Pappas, Theoni. *More Joy of Mathematics: Exploring Mathematics All Around You*. San Carlos, CA: Wide World, 1991.

Pappas, Theoni. *The Joy of Mathematics: Discovering Mathematics All Around You*. (2nd ed.). San Carlos, CA: Wide World, 1993.

Posamentier, Alfred S. *101+ Great Ideas for Introducing Key Concepts in Mathematics: A Resource for Secondary School Teachers*. Thousand Oaks, CA: Corwin Press, 2006.

Posamentier, Alfred S., and Daniel Jaye. *What Successful Math Teachers Do, Grades 6–12: 79 Research-Based Strategies for the Standards-Based Classroom*. Thousand Oaks, CA: Corwin Press, 2005.

Posamentier, Alfred S., Daniel Jaye, and Stephen Krulik. *Exemplary Practices for Secondary Math Teachers*. Alexandria, VA: Association for Supervision and Curriculum Development, 2007.

Principles and Standards for School Mathematics. Reston, VA: National Council of Teachers of Mathematics, 2000.

Reimer, Luetta, and Wilbert Reimer. *Mathematicians Are People, Too: Stories from the Lives of Great Mathematicians* (Vol. I). Palo Alto, CA: Dale Seymour, 1990.

Reimer, Luetta, and Wilbert Reimer. *Mathematicians Are People, Too: Stories from the Lives of Great Mathematicians* (Vol. II). Palo Alto, CA: Dale Seymour, 1993.

Secada, Walter G., Elizabeth Fennema, and Lisa Byrd Adajian (eds.). *New Directions for Equity in Mathematics Education.* New York: Cambridge University Press, 1995.

Silver, Harvey F., John R. Brunsting, and Terry Walsh. *Math Tools, Grades 3–12: 64 Ways to Differentiate Instruction and Increase Student Engagement.* Thousand Oaks, CA: Corwin Press, 2007.

Smith, Sanderson M. *Great Ideas for Teaching Math.* Portland, ME: Walch, 1990.

Smith, Sanderson M. *Agnesi to Zeno: Over 100 Vignettes from the History of Math.* Berkeley, CA: Key Curriculum Press, 1996.

Sobel, Max A., and Evan M. Maletsky. *Teaching Mathematics: A Sourcebook of Aids, Activities, and Strategies* (3rd ed.). Boston: Allyn & Bacon, 1998.

Sousa, David A. *How the Brain Learns Mathematics.* Thousand Oaks, CA: Corwin Press, 2007.

Stein, Sherman. *Mathematics: The Man-Made Universe* (3rd ed.). New York: Dover, 2000.

Usiskin, Zalman, Anthony L. Peressini, Elena Marchisotto, and Dick Stanley. *Mathematics for High School Teachers: An Advanced Perspective.* Upper Saddle River, NJ: Prentice Hall, 2002.

Van de Walle, John, Karen S. Karp, and Jennifer M. Bay Williams. *Elementary and Middle School Mathematics: Teaching Developmentally* (7th ed.). Boston: Allyn & Bacon, 2009

Zaslavsky, Claudia. *The Multicultural Math Classroom: Bringing in the World.* Portsmouth, NH: Heinemann, 1995.

Index

 How to Use the CD

System Requirements

PC with Microsoft Windows 2003 or later
   Mac with Apple OS version 10.1 or later

Using the CD With Windows

To view the items located on the CD, follow these steps:

1. Insert the CD into your computer's CD-ROM drive.
2. A window appears with the following options:

    Contents: Allows you to view the files included on the CD.

    Software: Allows you to install useful software from the CD.

    Links: Displays a hyperlinked page of websites.

    Authors: Displays a page with information about the author(s).

    Contact Us: Displays a page with information on contacting the publisher or author.

    Help: Displays a page with information on using the CD.

    Exit: Closes the interface window.

If you do not have autorun enabled, or if the autorun window does not appear, follow these steps to access the CD:

1. Click Start → Run.
2. In the dialog box that appears, type d:\start.exe, where d is the letter of your CD-ROM drive. This brings up the autorun window described in the preceding set of steps.
3. Choose the desired option from the menu. (See Step 2 in the preceding list for a description of these options.)

Note: to check a box in one of the documents on your computer, double-click on the box. Choose the option "Checked" under "Default value" in the dialog box that appears, then click "Okay."

In Case of Trouble

If you experience difficulty using the CD, please follow these steps:

1. Make sure your hardware and systems configurations conform to the systems requirements noted under "System Requirements" above.
2. Review the installation procedure for your type of hardware and operating system. It is possible to reinstall the software if necessary.

To speak with someone in Product Technical Support, call 800-762-2974 or 317-572-3994 Monday through Friday from 8:30 a.m. to 5:00 p.m. EST. You can also contact Product Technical Support and get support information through our website at www.wiley.com/techsupport.

Before calling or writing, please have the following information available:

- Type of computer and operating system.
- Any error messages displayed.
- Complete description of the problem.

It is best if you are sitting at your computer when making the call.